# Springer Tracts in Electrical and Electronics Engineering

Springer Tracts in Electrical and Electronics Engineering (STEEE) publishes the latest developments in Electrical and Electronics Engineering - quickly, informally and with high quality. The intent is to cover all the main branches of electrical and electronics engineering, both theoretical and applied, including:

- Signal, Speech and Image Processing
- Speech and Audio Processing
- Image Processing
- Human-Machine Interfaces
- Digital and Analog Signal Processing
- Microwaves, RF Engineering and Optical Communications
- Electronics and Microelectronics, Instrumentation
- Electronic Circuits and Systems
- Embedded Systems
- Electronics Design and Verification
- Cyber-Physical Systems
- Electrical Power Engineering
- Power Electronics
- Photovoltaics
- Energy Grids and Networks
- Electrical Machines
- Control, Robotics, Automation
- Robotic Engineering
- Mechatronics
- Control and Systems Theory
- Automation
- Communications Engineering, Networks
- Wireless and Mobile Communication
- Internet of Things
- Computer Networks

Within the scope of the series are monographs, professional books or graduate text-books, edited volumes as well as outstanding PhD theses and books purposely devoted to support education in electrical and electronics engineering at graduate and post-graduate levels.

**Review Process**

The proposal for each volume is reviewed by the main editor and/or the advisory board. The books of this series are reviewed in a single blind peer review process.

**Ethics Statement** for this series can be found in the Springer standard guidelines here https://www.springer.com/us/authors-editors/journal-author/journal-author-helpdesk/before-you-start/before-you-start/1330#c14214.

Ravindra Munje · Balasaheb Patre · Weidong Zhang

# Multi-stage Feedback Control Design for Multi-time-scale System

Springer

Ravindra Munje
Department of Electrical Engineering
K. K. Wagh Institute of Engineering
Education and Research
Nashik, Maharashtra, India

Balasaheb Patre
Department of Instrumentation Engineering
SGGS Institute of Engineering
and Technology
Nanded, Maharashtra, India

Weidong Zhang
Department of Automation
Shanghai Jiao Tong University
Shanghai, China

ISSN 2731-4200 ISSN 2731-4219 (electronic)
Springer Tracts in Electrical and Electronics Engineering
ISBN 978-981-97-4591-3 ISBN 978-981-97-4592-0 (eBook)
https://doi.org/10.1007/978-981-97-4592-0



This Springer imprint is published by the registered company Springer Nature Singapore Pte Ltd.
The registered company address is: 152 Beach Road, #21-01/04 Gateway East, Singapore 189721, Singapore

*To our families, parents, teachers, mentors, and friends, whose love, support, guidance, wisdom, and companionship have been the cornerstones of our journey. This book stands as a testament to your unwavering encouragement and inspiration.*

# Foreword I

It is a pleasure to write a Foreword for the monograph written by Ravindra Munje and his coworkers Prof. Balasaheb Patre and Prof. Weidong Zhang. I have been following Ravindra's research work and achievements with admiration. Although I have not taught or guided him directly as a student, he has been in constant touch with me to get new ideas and insights for his work. Dr. Patre and Dr. Zhang are both successfully leading the respective research groups and taking the control community to the next level. Being in the field of control engineering for several decades I have seen very few people who draw inspiration from others. Ravindra is one of them.

If we take a review, we can see that the mathematical theory of singularly perturbed and time-scale systems, also known as the theory of differential equations with small parameters multiplying certain derivatives, originated at the beginning of the 1950s and gained its maturity during the 1960s and 1970s. One of the most important results in the mathematical theory of linear singularly perturbed systems is the development of the Chang transformation, which facilitates the exact decomposition of singularly perturbed linear systems into pure-slow and pure-fast subsystems. Singularly perturbed control systems became an extensive subject of research by the end of the 1960s and during the 1970s in the papers published by P. Kokotovic and his graduate students, among whom P. Sannuti, J. Chow, H. Khalil, and D. Young were the most productive. A large number of journal papers on singularly perturbed control systems were published during the 1970s, 1980s, and 1990s in both mathematics and engineering. The approaches were taken in engineering during the 1970s and 1980s and applied to various systems for controller design purposes.

The presentation of the monograph is based on the recent research work of Ravindra and his coworkers. This monograph represents controller designs based on two-stage or three-stage decomposition techniques for continuous-time and discrete-time systems. The multi-stage feedback controllers are then studied using multi-stage observers. The monograph devises novel methods whose core result seems to be the reduction of online and offline computations, reduction in design complexity, and improvement in performance. It should be pointed out that some related problems remain unsolved, especially multirate observer designs using two-stage or three-stage decomposition. These problems are identified in the monograph as open problems for future research.

Each chapter is organized to represent an independent controller so that the readers interested in a particular method can find complete information within the particular chapter. The monograph demonstrates theoretical results on practical applications using examples from power plants.

One may say that multi-stage feedback control designs for multi-time-scale systems are moving into a new phase of control engineering. Ravindra, his coworkers, and the editors of this monograph deserve thanks for adding a volume to this re-invigorated branch of control system engineering.

September 2024

Zoran Gajic
Professor
Department of Electrical and Computer Engineering
Rutgers, The State University of New Jersey
Piscataway, NJ, USA

# Foreword II

It is indeed a privilege and pleasure for me to write the Foreword for a monograph written by Ravindra and his co-authors Prof. Balasaheb Patre and Prof. Weidong Zhang. I have known Ravindra since January 2013, when we met for the first time at the Indian Institute of Technology, Bombay. Balasaheb Patre is my early Ph.D. student. In fact, Dr. Patre introduced Ravindra to me during the QIP Short Term Course organized by me at IIT Bombay. Although not met Prof. Weidong Zhang personally, I have been following his work. I earnestly hope that this novel monograph on Multi-stage Feedback Control Designs for Multi-time Scale systems gets wide and wholehearted acceptance from the professional control engineering community consisting of R&D Labs and educational institutes, practicing engineers, university professors, and even graduate students. I take this opportunity to congratulate the authors for writing such an excellent monograph.

Widely spaced groups of eigenvalues in multi-time-scale systems result in ill-conditioned system matrices. This creates difficulty in developing straightforward feedback controllers and reduced- or full-order observers. As a result, the design is completed in multiple stages, namely two-stage design for the two-time-scale system and three-stage design for the three-time-scale system. Further, for the digital implementation, sampling of the states with the same sampling interval does not give the desired performance. Hence, the sampling of the multi-time-scale system is done with different sampling periods. Various control strategies for multi-stage and multirate feedback controls are presented in this monograph.

Chapter 1 provides a review of the literature on multi-stage feedback control designs for multi-time-scale continuous-time and discrete-time systems. Chapters 2–4 present the design of multi-stage state feedback controllers in the continuous-time domain. In this, Chaps. 3 and 4 also demonstrate the applicability of the suggested controllers for nonlinear models of power plants using multi-stage reduced- and full-order observers. Chapters 5–7 discuss multirate state feedback controllers using a time-scale decomposition technique. These designs are also applied to the linear models of power plants using sequential multi-stage observers. Finally, Chap. 8 gives a comparison of these controllers from the design point of view.

A reader of this monograph should have a general background in power plant modeling, modern control theory, and singular perturbation methodologies, including some knowledge of linear algebra. This monograph is useful primarily for research scholars.

Finally, I wish a grand success to this endeavor of authors and I am sure that this monograph will fulfill the needs of numerous professors and research students of modern control theory and will be a useful contribution to multi-stage feedback control design for multi-time-scale systems.

September 2024

Bijnan Bandyopadhyay
Professor
Department of Systems and Control Engineering
Indian Institute of Technology Bombay
Mumbai, Maharashtra, India

# Preface

Multi-time-scale systems have widely separated clusters of eigenvalues, making the system matrices ill-conditioned. Due to this the direct design of a feedback controller and reduced- or full-order observers for such systems is a challenging task. Hence, the feedback controllers and observers are designed in multiple design stages, i.e., two-stage design for the two-time-scale system and three-stage design for the three-time-scale system. In this monograph, a multi-stage feedback controller and observer designs are proposed for the three-time-scale system to reduce offline and online computational efforts. The applicability of these methods is demonstrated by simulating practical systems like a coal-fired power plant and an advanced heavy water reactor.

This monograph also discusses the design of multi-stage feedback controllers in discrete-time. The design of state feedback control for a multi-time-scale system with a lower sampling rate fails to capture information in the fastest (very fast) and fast states. While sampling at a higher rate increases computation time. Thus, the use of single-rate sampling for such systems is unsuitable. Therefore, multirate state feedback controllers are designed for the multi-time-scale system, in which slow, fast, and very fast states are sampled at different sampling rates. As slow, fast, and very fast states are the decoupled and internal states of the original system, these are estimated by the sequential multi-stage observers.

Starting with the brief literature review in the first chapter, the various designs listed below are presented in separate chapters. The designs presented in this monograph are as follows:

1. Design of a three-stage feedback controller for a continuous-time three-time-scale system using the two-time-scale approach.
2. Design of two-stage and three-stage reduced- and full-order observers for the estimating states of the advanced heavy water reactor, which are fed to the feedback controller for spatial stabilization.
3. Design of the three-stage feedback control for the coal-fired power plant using the internal model control principle for achieving robust tracking and disturbance rejection performance.

4. Design of two-stage multirate state feedback control for the two-time-scale system depending upon the sampling sequence.
5. Design of two-stage multirate state feedback control for the two-time-scale system in which the decoupled states are estimated by the two-stage sequential observer.
6. Design of a three-stage multirate state feedback control scheme for the three-time-scale system in which the decoupled states are determined by the three-stage sequential observer.

These methods result in considerable savings in online and offline computations, reduction in design complexity, and improvement in robustness and closed-loop performance. A comparative summary of the control designs is presented in the last chapter along with future areas of research.

This monograph can be a stepping stone for scientists, researchers, and mathematicians working on designing controllers for multi-time-scale systems both in the continuous-time domain as well as the discrete-time domain. The authors would like to thank IEEE for allowing them to reuse their protected content in this monograph. The authors express sincere thanks to the National Natural Science Foundation of China (NSFC), China for funding the research project entitled 'Multi-stage Feedback Control Design for Multi-time-scale System for Continuous and Discrete-Time Systems' (Project ID: 61750110524). The authors also would like to say thanks to Dr. S. R. Shimjith for developing a multi-time-scale model of the Advanced Heavy Water Reactor for which the various controllers are developed under the grant 2009/36/102-BRNS/3284 given by the Board of Research in Nuclear Sciences, Department of Atomic Energy, Government of India to Dr. B. M. Patre. The same mathematical model has been used in one of the chapters of this monograph.

The authors would like to express their deep sense of gratitude to their parents and teachers who have made them competent enough to write this monograph. The authors wish to thank many individuals who helped them directly or indirectly in completing this monograph, particularly Dr. Danying Gu, Dr. Shuyi Lin, Dr. Guoqing Zhang, and Mr. Ranvir Desai. Finally, the authors wish to acknowledge the support, patience, and love of their spouses and children during the preparation of this monograph.

Nashik, India — Ravindra Munje
Nanded, India — Balasaheb Patre
Shanghai, China — Weidong Zhang
September 2024

# Contents

# About the Authors

**Dr. Ravindra Munje** holds a B.E., M.E., and Ph.D. degrees from Mumbai University in 2005, Pune University in 2009, and Swami Ramanand Teerth Marathwada University in 2015 respectively. He completed his post-doctoral fellowship at Shanghai Jiao Tong University in Shanghai from June 2017 to May 2019. During his fellowship, he received research funds of approximately 15 lakh rupees from the National Natural Science Foundation of China as an International Young Scientist. Currently, Dr. Munje is a Professor and head of the Electrical Engineering Department at K. K. Wagh Institute of Engineering Education and Research in Nashik, India. He also holds the position of Dean (Research and Development) at the institute. He has authored a book and about 60 refereed papers, with his research interests focusing on modeling and control of large-scale systems using sliding mode and multi-rate output feedback. He has also supervised eight PG students and is currently guiding four Ph.D. students. Dr. Munje has been recognized with several awards, including the Promising Engineer Award in 2016, the Outstanding Researcher Award in 2018, and the Best Faculty Award in 2019. Recently, he was awarded the prestigious 'Best Teacher Award' of Savitribai Phule Pune University in February 2024. He is a senior member of IEEE, a member of IET, and a life member of Indian Society for Technical Education (ISTE).

**Dr. Balasaheb Patre** completed a Bachelor of Instrumentation Engineering in 1986 and, a Masters of Instrumentation Engineering in 1990, both from SGGS Institute of Engineering and Technology, Vishnupuri, Nanded, and Ph.D. from Indian Institute of Technology (IIT), Bombay in 1998. Currently, he is working as a Professor at the Department of Instrumentation Engineering, SGGS Institute of Engineering and Technology, Vishnupuri, Nanded, Maharashtra. He has published more than 80 International Journal Papers, more than 100 National and International Conference papers, six books, and five book chapters. He is a reviewer for 45 international (SCI/SCIE indexed) peer-reviewed journals. He has completed sponsored research projects from DST, AICTE, BARC, BRNS, and NRB. He has so far supervised 22 students for Ph.D. and five students are currently pursuing Ph.D. under his guidance in the areas of systems and control engineering in general and robust control, nuclear reactor control, sliding mode control and applications, and process control in particular. Dr. Patre is a senior member of IEEE, a Fellow of the Institution of Engineers (India), a Member of IETE (India), and a life member of ISTE.

**Weidong Zhang** received his B.S., M.S., and Ph.D. degrees from Zhejiang University, China, in 1990, 1993, and 1996, respectively, and then worked as a Postdoctoral Fellow at Shanghai Jiaotong University. He joined Shanghai Jiaotong University in 1998 as an Associate Professor and has been a Full Professor since 1999. From 2003 to 2004 he worked at the University of Stuttgart, Germany, as an Alexander von Humboldt Fellow. From 2007 to 2008 he worked at Princeton University, USA, as a Visiting Professor. From 2013 to 2017 he serviced as Deputy Dean of the Department of Automation, Shanghai Jiaotong University. He serving as a part-time professor at Hainan University from 2021. He is currently the Director of the Engineering Research Center of Marine Automation, Shanghai Municipal Education Commission, China. His research interests include control theory, machine learning theory, and their applications in industry and robots. He is the author of more than 300 papers and 2 books and has been recognized as an Elsevier most cited researcher.

# Acronyms

| | |
|---|---|
| AHWR | Advanced Heavy Water Reactor |
| ARX | Auto Regressive with External Input |
| CFPP | Coal-Fired Power Plant |
| DEB | Direct Energy Balance |
| FLOP | Floating Point Operation |
| GW | Giga Watt |
| HOSMO | Higher Order Sliding Mode Observer |
| IAE | Integral Absolute Error |
| ISE | Integral Square Error |
| LQR | Linear Quadratic Regulator |
| MFC | Multirate Feedback Control |
| MPa | Mega Pascal |
| MRSF | Multirate State Feedback |
| NSFC | National Science Foundation of China |
| MW | Mega Watt |
| PI | Proportional Integral |
| PKM | Point Kinetic Model |
| RR | Regulating Rod |
| SMO | Sliding Mode Observer |

# Symbols

## Operations

| | |
|---|---|
| Matrix | Bold upper case |
| Vector | Bold lower case |
| $\mathbf{0}$ | Zero vector or zero matrix |
| $\mathbf{I}_n$ | Identity matrix of dimension $n$ |
| $\mathbf{A}^T$ | Transpose of matrix $\mathbf{A}$ |
| $\mathbf{A}^{-1}$ | Inverse of square matrix $\mathbf{A}$ |
| $\mathbf{A}^{-T}$ | Inverse of transposed square matrix $\mathbf{A}$ |
| $\psi(\mathbf{A})$ | Eigenvalues of square matrix $\mathbf{A}$ |
| $max(\cdot)$ | Maximum of function |
| $min(\cdot)$ | Minimum of function |
| $rank\{\mathbf{A}\}$ | Rank of matrix $\mathbf{A}$ |
| $det(\mathbf{A})$ | Determinant of matrix $\mathbf{A}$ |
| $\Re^n$ | Real space of dimension $n$ |
| $\mathbf{z} \in \Re^n$ | $\mathbf{z}$ Belongs to real space of dimension $n$ |
| $\Re e\{\mathbf{x}\}$ | Real part of $\mathbf{x}$ |
| $\Im m\{\mathbf{x}\}$ | Imaginary part of $\mathbf{x}$ |
| $\|\cdot\|$ | Modulus of function |
| $\sum_{i=1}^{N}$ | Summation $i = 1$ to $N$ |
| $\frac{d\mathbf{z}}{dt} = \mathbf{z}$ | Time derivative of $\mathbf{z}$ |

## Notations

| | |
|---|---|
| $\mathbf{A}, \mathbf{B}, \mathbf{C}$ | State, input, and output matrices of a linear continuous-time system |
| $\mathbf{F}$ | State feedback gain matrix of discrete-time system |
| $\mathbf{K}$ | State feedback gain matrix of continuous-time system |
| $\mathbf{L}$ | Observer gain matrix |

| | |
|---|---|
| $N, N_1, N_2$ | Sampling intervals |
| $\mathbf{M}$, $\mathbf{P}$, $\mathbf{T}$, $\mathbf{U}$ | Linear state transformation matrices |
| $T_s$, $T_c$ | Sampling periods |
| $j$ | Complex number |
| $m$ | Number of inputs |
| $n$ | Number of states or system order |
| $p$ | Number of outputs |
| $t$ | Time, s |
| $\mathbf{u}$ | Input vector |
| $\mathbf{y}$ | Output vector |
| $\mathbf{z}$ | State vector |
| $\mathbf{\Phi}$, $\mathbf{\Gamma}$ | State and input matrices of a linear discrete-time system |
| $\tau$, $\Delta$, $\delta$ | Sampling periods |
| $\alpha$, $\beta$ | Controllability and Observability indices |
| $\varepsilon$, $\mu$ | Singular perturbation parameters |
| $\psi_i$ | $i$th eigenvalue |

# List of Figures

# List of Tables

# Chapter 1
# Introduction

**Abstract** This chapter offers a brief survey of the state-of-the-art literature available on the multi-stage feedback control designs for both continuous-time and discrete-time systems. For continuous-time systems, discussions are given on two-stage and three-stage designs of feedback controllers and observers. Whereas for discrete-time systems, in addition to what is stated for continuous-time systems, thoughts are given on multirate state feedback designs. Later the motivation for writing this monograph is highlighted. The chapter ends with the organization of the monograph.

**Keywords** Multi-stage feedback design · Two-stage design · Three-stage design · Singular perturbation · Time-scales

## 1.1 Introduction

The analysis and control of large-scale systems become more difficult due to the system's high-order nature and coupled dynamics which often vary with different speeds. These types of systems are comprehensively studied by singular perturbation and time-scale methods (Naidu 1988; Kokotovic et al. 1999; Gajic and Lim 2001). Singular perturbations have long been known to exist in most classical and contemporary control schemes based on reduced-order models. This knowledge has prompted the development of time-scale methods for a variety of applications, using state feedback, output feedback, filter, and observer designs. Over the period, excellent overviews and surveys have been published for the periods up to 1976 (Kokotovic et al. 1976), 1976–1983 (Saksena et al. 1984), 1984–2001 (Zhang et al. 2014), and 2002–2012 (Naidu 2002). Surveys on singular perturbation and time-scale methods for modeling physical systems (Syrcos and Sannuti 1983) and discrete systems (Naidu et al. 1987) are also documented. The fact that there is such a large body of literature on singular perturbation is owing both to the theoretical interest in the subject and to the practical relevance of this research for issues in control engineering. It has been shown to be a useful tool for control design.

Dynamical systems with multi-time scales arise naturally in many areas. The peculiar nature of these systems provides the principal motivation for this research.

 
R. Munje et al., *Multi-stage Feedback Control Design for Multi-time-scale System*,
Springer Tracts in Electrical and Electronics Engineering,
https://doi.org/10.1007/978-981-97-4592-0_1

Multi-time-scale systems typically have small time constants or other 'parasitic' factors, which are often disregarded in simplified modeling. If one uses a simplified model to design a control strategy, the desired performance may not be achieved or the system may be unstable. If the small quantities are taken into account the system's dynamics as a whole can be investigated, but its order can be raised, and the calculations needed to develop a controller may become time-consuming and even inaccurate. Alternatively, to simplify the design process, traditional control schemes for ill-conditioned systems frequently overlook the fast dynamics. These controller designs often include a trade-off between the need for computation and the loss of system data. The primary objective of the multi-time-scale technique is to reduce the high dimensionality and ill-conditioning brought on by the combination of slow and fast dynamic modes. When two-time-scale systems are broken down into separate slow and fast subsystems, for instance, it is suggested that separate slow and fast controls be created for each subsystem before being integrated to provide a 'composite' control for the original system. Only the slow, 'reduced', control is employed if the fast control is not required. Numerous two-time-scale designs have been inspired by these concepts. However, in many multi-time-scale systems with hundreds or thousands of state variables, the two-time-scale approach does not work because the various submatrices produced by the two-stage decomposition are just as poorly conditioned as the matrices of the original system. In that case, one has to go for the multi-stage decomposition. Another motivation for the multi-stage feedback control design is that the errors in one stage do not transfer to the next stage. Hence, more robustness can be achieved compared to a two-stage design. A brief literature review of the existing methodologies for feedback control design for linear multi-time-scale continuous-time and discrete-time systems is given below.

## 1.2 Literature Review

The review is not meant to be thorough, but it does include the majority of the sources that inspired or relate directly to the development of this monograph.

### *1.2.1 Continuous-Time Systems*

Straightforward feedback control designs for multi-time-scale systems often lack accuracy. Considering multi-time-scale systems, slow and fast modes of the eigenvalue exist in two-time-scale systems, whereas slow, fast, and very fast modes exist in three-time-scale systems. Controllers for these types of systems are efficiently constructed by dividing the original system into lower-order subsystems. The lower-order subsystems are obtained by the quasi-steady-state approach and direct block-diagonalization. The quasi-steady-state technique is effective for very small perturbation parameters. When the quasi-steady-state technique is applied, the eigenvalues of the slow and fast subsystems are no longer in the same location as the eigenvalues of the full-order system if the perturbation parameter is not zero. As a result, a two- or three-stage block-diagonalization procedure can be used. Accurate subsystem decoupling is achieved using this technique. Control law synthesis for such systems can be performed on each separate subsystem, and then the results are combined to produce a composite feedback control for the original system.

In the singular perturbation approach to feedback control design, the interaction of these modes is taken advantage of by decomposing the original ill-conditioned system into lower-order subsystems (Naidu 1988). The singularly perturbed two-time-scale nature of the system motivates the transformation of the main system into the slow and fast subsystems. Such a transformation for linear systems is constructed by Chang (1972). The feedback design for each subsystem can then proceed, and the results are merged to create an overall feedback control for the main system. A feedback control design using a quasi-steady-state method for the two-time-scale system is presented in Chow and Kokotovic (1976). In this, to move the eigenvalues of the original system sufficiently close to the desired positions, the composite feedback gain is derived by combining the results of two gain matrices intended for the placement of slow and fast eigenvalues separately. A similar approach for a three-time-scale system is recommended in Shimjith et al. (2011) for a large-scale system. As there is no need to apply matrix transformations to the original system and also no need to solve iterative algebraic equations, the eigenvalue assignment approach with the methods of Chow and Kokotovic (1976); Shimjith et al. (2011) is much simpler. However, because of the approximation, the subsystem eigenvalues of the closed-loop system are not placed precisely.

The approximation of the quasi-steady-state method can be overcome by a two-stage feedback control design, based on explicit invertible block-diagonal transformation (Phillips 1980a). In Nguyen and Naidu (2012), the application of wind energy conversion system is taken to demonstrate the applicability of Phillips (1980a). In this case, using the linear quadratic regulator (LQR) theory, independent lower-order optimal controllers are created for the slow and fast subsystems of the original system. To demonstrate the superiority of the proposed method in terms of separation of dynamics and less computational effort, the reduced-order optimal control of independent subsystems is compared with the higher order optimal control of the original system. On similar lines, the control of unmanned aerial vehicles (Archibong and Naidu 2014) and analysis of electrical transmission lines in a smart grid (Jaison et al. 2015) are discussed using a two-time-scale singular perturbation approach. Simulations presented in Nguyen and Naidu (2012); Archibong and Naidu (2014); Jaison et al. (2015) show good agreement between the original linear system and the slow and fast subsystems. As an extension to Phillips (1980a), a simplified two-stage design is presented in Radisavljevic-Gajic and Rose (2014). In Radisavljevic-Gajic and Rose (2014), the full-order system is initially transformed into an upper triangular form, and the first stage of design is achieved. Then, the system is modified to a lower triangular form and the second stage is completed. In contrast to Phillips (1980a) where three transformations are required, in Radisavljevic-Gajic and Rose

(2014) only two transformations are needed. Further, the choice of designing independent ‘slow’ and ‘fast’ controllers is preserved.

As a first step of designing a three-stage feedback controller for the three-time-scale system, direct block-diagonalization of the three-time-scale system into ‘slow’, ‘fast’, and ‘very fast’ modes, using linear transformation matrices is outlined in Naidu and Ravinder (1985). Furthermore, in Singh et al. (1986), a two-stage decomposition procedure (Phillips 1980a) of decoupling two-time-scale system into ‘slow’ and ‘fast’ subsystems is used to decouple the three-time-scale system into three lower-order subsystems. However, in both Naidu and Ravinder (1985) and Singh et al. (1986), the study is restricted only to autonomous systems, i.e., feedback control design aspects are not elaborated. In Kodra et al. (2016), a method for studying multi-time-scale singularly perturbed systems that maintain time-scale features is proposed. The method is based on the Schur decomposition. First, a controller design based on the assignment of eigenvalues is recommended, using the state vectors of each time scale as feedback. Second, a combined linear quadratic controller technique with eigenvalue assignment is investigated. It has been demonstrated that the combined design technique is flexible and can be more efficient for particular control issues.

In the recent past, Radisavljevic-Gajic et al. (2017), have reported a method for three-stage feedback control design for three-time-scale system with reduced computation. Instead of designing the feedback controls after block-diagonal transformation (Naidu and Ravinder 1985), in this, the original three-time-scale system is transformed into an upper triangular form by similarity transformations. Subsequently, controllers for very fast, fast, and slow subsystems are designed in a sequence. These are then integrated to create a comprehensive composite state feedback control for the initial system. In contrast to the regular three-stage design using block-diagonalization, which requires the solution of nine equations, the proposed feedback control design requires the solution of only six algebraic equations. Another approach for a three-time-scale system is documented in Munje et al. (2018). This technique transforms the system into a lower triangular form with two-time-scale similarity transformations, generating separate feedback controls for subsystems. This simplifies computations, reduces matrix calculations, and significantly cuts down computation time. Further, an integral state feedback controller, coupled with an observer is devised for enhanced control performance in Munje et al. (2020) using a three-time-scale approach. It ensures zero steady-state error, utilizes innovative two-time-scale separation to simplify design, and improves performance. It is demonstrated for a coal-fired power plant.

Similar to feedback control gain, observer gain for the multi-time-scale system can also be designed in the multiple stages (O’Reilly 1980; Shouse and Taylor 1993; Yoo and Gajic 2017, 2018). In O’Reilly (1980), full-order and in Shouse and Taylor (1993) reduced-order observers are presented for linear singularly perturbed continuous-time two-time-scale systems. But then again, the designs are carried out using the quasi-steady-state method. In Yoo and Gajic (2017), reduced-order observers are designed for the two-time-scale system by considering five different cases of available states, e.g., slow states, fast states, and some combinations of slow and fast states. While in Yoo and Gajic (2018), the full-order observer is designed for the same system using the methodology of Phillips (1980a). Here also, two schemes are discussed, namely sequential observation and parallel observation, in which slow and fast states are separately estimated. Design methods are well-explained by solving numerical examples. Furthermore, the study in Munje et al. (2018) introduces full and reduced-order observers with two-stage and three-stage designs for multi-time-scale systems. These innovative multi-stage observers simplify design and computation while maintaining flexibility for independent observer gain designs. These schemes are applied to the state estimation of the Advanced Heavy Water Reactor (AHWR) for enhancing spatial stabilization in conjunction with a feedback controller and for improving performance compared to using the controller without an observer.

### *1.2.2 Discrete-Time Systems*

The digital design of a feedback control system has gained much importance, due to the development of microelectronics and computing technology. This has sparked an interest in researching discrete-time systems with linear state feedback controllers. For the discrete-time systems with slow and fast states, feedback control design, however, becomes very challenging (Naidu et al. 1987). The reason being, the former states describe the long-term behavior of the system and later states express short-term behavior, i.e., eigenvalues of the system are distributed into a group near the unit circle and a group near the origin (Zhang et al. 2014). This type of eigenvalue clustering promotes the feedback control design in multiple stages.

Two-stage design of feedback control for a discrete-time system with slow and fast dynamic modes first appeared in the 1980s Phillips (1980b). In this, an explicit invertible block-diagonalization procedure, for obtaining reduced-order subsystems and subsequent composite feedback control design, is introduced. One more methodology based on quasi-steady-state modeling is outlined by Mahmoud (1983); Mahmoud et al. (1985). However, the purpose of placing subsystem eigenvalues at the preferred locations is not fully accomplished due to approximation in the quasi-steady-state method. Recently, simplified versions of a technique advised in Phillips (1980b) for the two-stage design of linear feedback controller are developed in Radisavljevic-Gajic (2015a, b). These techniques offer significant savings in online and offline computations. The difference in Radisavljevic-Gajic (2015a, b) is in the type of two-time-scale formulation used for two-stage design purposes. In Radisavljevic-Gajic (2015a) fast-time-scale formulation is considered while in Radisavljevic-Gajic (2015b) slow-time-scale formulation is operated. As far as the three-stage design of feedback control for the three-time-scale system is concerned, a variety of three-time-scale discrete-time models are suggested in Radisavljevic-Gajic et al. (2019). However, a well-accepted explicit singularly perturbed three-time-scale model of the linear discrete-time system is not available.

In general, the singular perturbation parameter need not appear explicitly in the system. Therefore, it might not always be possible to apply solitary perturbation approaches. Instead, the use of similarity transformations to derive decoupled subsystems could be a more promising method (Ladde et al. 1985). Using this approach, three-stage designs for the discrete-time system are presented in Naidu (1988) for the system with slow, fast, and very fast eigenvalues. An application of this to the AHWR is demonstrated in Shimjith et al. (2013); Munje et al. (2016).

It must be noted that during a short transient period, the response of fast or very fast states is worth observing. Later, slow states come into the picture to control the system's behavior. When designing feedback controls for these systems, if a longer sample interval is used, it results in information loss in fast-varying states and if a shorter sample interval is used, it significantly increases the time required for online computing. As a result, it is not a good idea to design feedback control at a single sample rate. Conversely, using a multirate sampling strategy (Glasson 2008) designer can accommodate multiple sampling rates for the group of states, rather than a single sampling rate, i.e., fast-varying states can be sampled faster than slowly varying states. The performance of a closed-loop control system can be improved as a result.

The conceptual background for investigating multirate feedback control for the systems with slow and fast-varying modes is established in Litkouhi and Khalil (1985). In this, two multirate sampling schemes have been suggested. Further, Kando and Iwazumi (1986) recommended the design of optimal regulators via multirate sampling for the discrete two-time-scale system. Lennarston (1989) demonstrated that the time-varying disturbance, not explored in Litkouhi and Khalil (1985); Kando and Iwazumi (1986), can be efficiently taken care of by adding immeasurable states in the existing model and also revealed that these states can be determined by an estimator. A different study on disturbance rejection is reported in Yamamoto et al. (2017), in which disturbance signals below and above Nyquist frequency are treated. Apart from disturbance rejection, multirate sampling is also employed to resolve the issue of unstable zeros of the discrete-time system using the two-delay output and input controls (Mita et al. 1990) and multirate feedforward control (Fujimoto et al. 2001). The two-delay output control handles disturbances and parameter variations and in two-delay input control, an input observer is constructed to achieve loop transfer recovery. Even so, two-delay digital controls (Mita et al. 1990) result in large overshoot and oscillations in inter-sample points. Fujimoto et al. (2001) submitted a multirate feedforward controller to overcome this problem by optimizing inter-sample response. Although promising, works in Yamamoto et al. (2017); Mita et al. (1990); Fujimoto et al. (2001) are restricted to a general class of linear time-invariant systems, and no time-scale decoupling is dealt with.

State estimation problems (Lennarston 1989; Mita et al. 1990) are also solved by reporting multirate observers (Kando et al. 1998, 2000; Bidani and Djemai 2002; Ling and Kravaris 2017). In Kando et al. (1998, 2000) multirate full- and reduced-order observers are integrated with the multirate controller of Kando and Iwazumi (1986) and in Bidani and Djemai (2002) the multirate controller and observer are incorporated for the nonlinear two-time-scale system. Another multirate observer for a nonlinear autonomous system, based on a continuous-time design coupled with inter-sample predictors for the sampled measurements is presented in Ling and Kravaris (2017). But then, separate estimation of slow and fast states is not discussed. On the contrary, even if the techniques in Kando et al. (1998, 2000), Bidani and Djemai (2002) estimate slow and fast states separately, they use the quasi-steady-state method for decoupling the two-time-scale system, which is accompanied by the approximation, as mentioned before.

In Munje and Zhang (2021), a multirate state feedback control strategy is introduced for a two-time-scale system. It involves obtaining a block-triangular form of the continuous system, discretizing it step by step, and designing feedback controls at different sampling rates. A two-stage observer is proposed to estimate the slow and fast states, and the applicability of the method is demonstrated with a numerical example. In Munje and Patre (2023), two multirate sampling approaches are discussed. The first approach starts with the fast subsystem and then samples the slow subsystem, while the second approach begins with the slow subsystem before sampling fast ones. The feedback control design is done in two stages. The first stage is initiated after the first subsystem sampling and the second stage after sampling the second subsystem. The study compares and applies these methods in a numerical example.

## 1.3 Motivation

As said earlier, the main motivation is the presence of the multi-time-scale property in the physical system, i.e., the eigenvalues of the system are grouped into clusters as slow, fast, and very fast. The decoupling of a multi-time-scale system into lower-order subsystems provides the scope to have a better understanding of the different dynamics of the system. Having stable fast and very fast subsystems, it would be interesting to examine the asymptotically stable closed-loop performance of the system with the reduced-order controllers. Further, with the full-order (global) controller, obtained from the local feedback controllers designed for the individual subsystems, achieving an improved and optimized closed-loop performance would be challenging.

The digital architecture of the equivalent controllers would enable the usage of various sampling rates for the fast (faster sampling rates) and slow (slower sampling rates) controllers. A single sample rate would otherwise be necessary for the digital controller of the entire system without the multi-stage design, which could be computationally expensive or cause performance to degrade. One more motivation for the digital implementation of the feedback controller for the three-time-scale system is to investigate the relationship between the sampling periods and singular perturbation parameters of the system. For multirate sampling, estimation of the decoupled states with the sequential or parallel observer would play an important role. Otherwise, the multirate sampling design methods would be restricted to simulation studies only.

Since the designs are completed in separate stages, any mistakes made at a later level wound not have an impact on the accuracy of the designs at a previous stage. Therefore, adapting the approach to a multi-stage design is an intriguing study issue.

In addition to the already-demonstrated simplifications, the design would be crucial in enhancing system robustness; therefore, the initial stages of design should be carried out with extremely high accuracy, while later stages may allow for some errors because they will only affect specific subsystems and not the entire system.

## 1.4 Organization of the Monograph

The monograph is organized as follows:

1. This chapter provides an introduction, a brief review of literature pertinent to this work and motivation.
2. Chapter 2 presents a three-stage feedback control design for a three-time-scale system by using two-time-scale similarity transformations.
3. Chapter 3 proposes three-stage and two-stage designs of the full- and reduced-order observers for the three-time-scale system with its application to the AHWR.
4. Chapter 4 demonstrates an extension of the three-stage feedback control design to the system with integral action for achieving robust tracking performance of the CFPP. The states of CFPP are estimated using a full-order observer.
5. Chapter 5 recommends the feedback control designs for a linear time-invariant system, having slow and fast eigenvalue clusters, by multirate sampling of states in two stages.
6. Chapter 6 suggests a novel multirate state feedback control scheme, in which states are estimated by a sequential observer, for the linear time-invariant two-time-scale system.
7. Chapter 7 presents a three-stage multirate state feedback control strategy for the linear time-invariant three-time-scale system, where states are determined by the three-stage observer.
8. Chapter 8 provides overall conclusions and future directions.

## References

Archibong E, Naidu DS (2014) Time scale analysis and synthesis for unmanned aerial vehicles. In: International Symposium on Resilient Control Systems, pp 1–6

Bidani M, Djemai M (2002) A multirate digital control via a discrete-time observer for nonlinear singularly perturbed continuous-time systems. Int J Control 75:591–613

Chang KW (1972) Singular perturbations of a general boundary value problem. SIAM J Math Anal 3:520–526

Chow JH, Kokotovic PV (1976) Eigenvalue placement in two-time-scale systems. Proc IFAC Symp Large Scale Syst Theory Appl 9:321–326

Fujimoto H, Hori Y, Kawamura A (2001) Perfect tracking control based on multirate feedforward control with generalized sampling periods. IEEE Trans Ind Electron 48:636–644

Gajic Z, Lim M-T (2001) Optimal control of singularly perturbed linear systems and applications: high accuracy techniques. Marcel Dekker, New York, USA

Glasson D (2008) Development and applications of multirate digital control [25 Years Ago]. IEEE Control Syst Mag 28:13–15
Jaison S, Gentle JP, Naidu DS (2015) Time scale analysis and synthesis for electrical transmission lines in a smart grid. In: IEEE Conference Technologies for Sustainability, pp 103–108
Kando H, Iwazumi T (1986) Multirate digital control design of an optimal regulator via singular perturbation theory. Int J Control 44:1555–1578
Kando H, Ukai H (1998) Design of reduced-order multirate observers and stabilizing multirate controllers via singular perturbation theory. Optim Control Appl Meth 19:435–445
Kando H, Uhai H, Morita Y (2000) Design of multirate observers and multirate control systems. Int J Syst Sci 31:1021–1030
Kodra K, Zhong N, Gajic Z (2016) Multi-time-scale systems control via use of combined controllers. In: European control conference, pp 2638–2643
Kokotovic PV, Khalil HK, O'Reilly J (1999) Singular perturbation methods in control: analysis and design. SIAM Publishers, Philadelphia
Kokotovic PV, O'Malley RE, Sannuti P (1976) Singular perturbation and order reduction in control theory-an overview. Automatica 12:123–132
Ladde GS, Rajalakshmi SG (1985) Diagonalization and stability of multi-time- scale singularly perturbed linear systems. Appl Math Comp 16:115–140
Lennarston B (1989) Multirate sampled-data control of two-time-scale systems. IEEE Trans Auto Control 34:642–644
Ling C, Kravaris C (2017) Multirate sampled-data observers based on a continuous-time design. In: IEEE annual conference decision and control, pp 3664–3669
Litkouhi B, Khalil H (1985) Multirate and composite control of two-time-scale discrete-time systems. IEEE Trans Auto Control 30:645–651
Mahmoud MS (1983) Design of feedback controllers by two-stage methods. Appt Math Model 7:163–168
Mahmoud MS, Chen T, Singh MG (1985) On eigenvalue assignment in discrete systems with fast and slow modes. Int J Syst Sci 16:61–70
Mita T, Chida Y, Kaku Y, Numasato H (1990) Two-delay robust digital control and its applications-avoiding the problem on unstable limiting zeros. IEEE Trans Auto Control 35:962–970
Munje RK, Gu D, Xu B, Zhang W (2018) Feedback control design for three-time-scale system using two-time-scale approach. In: Proceedings Chinese control conference, Wuhan, pp 4982–4987
Munje R, Zhang WD (2021) Observer-based multirate feedback control design for two-time-scale system. Int J Autom Comput 18:1007–1016
Munje RK, Patre BM, Tiwari AP (2016) Discrete-time sliding mode control of advanced heavy water reactor. IEEE Trans Control Syst Technol 24:357–364
Munje RK, Gu D, Desai R, Patre BM, Zhang W (2018) Observer-based spatial control of advanced heavy water reactor using time-scale decoupling. IEEE Trans Nucl Sci 65:2756–2766
Munje RK, Lin S, Zhang G, Zhang W (2020) Observer-based output feedback integral control for coal-fired power plant: a three- time-scale perspective. IEEE Trans Control Syst Technol 28:601–608
Munje R, Patre B (2023) Two-stage multirate state feedback control designs for systems with slow and fast eigenvalue modes. Trans Inst Meas Control 46:1027–1037. https://doi.org/10.1177/01423312231187445
Naidu DS, Ravinder R (1985) On three time scale analysis. In: Proceedings IEEE conference on decision and control, pp 81–85
Naidu DS (2002) Singular perturbations and time scales in control theory and applications: overview. Dyn Contin, Discret Impuls Syst J 9:233–278
Naidu DS (1988) Singular perturbation methodology in control systems. Peter Peregrinus Ltd, London U.K
Naidu DS, Price DB, Hibey JL (1987) Singular perturbations and time-scales (SPaTS) in discrete control systems-an overview. Proc IEEE Conf Deci Control 26:2096–2103

Nguyen HM, Naidu DS (2012) Time scale analysis and control of wind energy conversion systems. In: International symposium on resilient control systems, pp 149–154
O'Reilly J (1980) Dynamical feedback control for a class of singularly perturbed linear systems using a full-order observer. Int J Control 31:1–10
Phillips R (1980a) A two-stage design of linear feedback controls. IEEE Trans Auto Control 25:1220–1223
Phillips RG (1980b) Reduced order modelling and control of two-time-scale discrete systems. Int J Control 31:765–780
Radisavljevic-Gajic V (2015a) A simplified two-stage design of linear discrete-time feedback controllers. ASME J Dyn Syst, Meas, Control 137:1–7
Radisavljevic-Gajic V (2015b) Two-stage feedback control design for a class of linear discrete-time systems with slow and fast modes. ASME J Dyn Syst, Meas, Control 137:1–6
Radisavljevic-Gajic V, Rose P (2014) A new two-stage design of feedback controllers for a hydrogen gas reformer. Int J Hydrog Energy 39:11738–11748
Radisavljevic-Gajic V, Milanovic M, Clayton G (2017) Three-stage feedback controller design with application to three-time-scale linear control system. ASME J Dyn Syst, Meas, Control 139:1–10
Radisavljevic-Gajic V, Milanovic M, Rose P (2019) Three-stage discrete-time feedback controller design. In: Multi-stage and multi-time scale feedback control of linear systems with applications to fuel cells. Springer Nature Switzerland AG, pp 97–108
Saksena VR, O'Reilly J, Kokotovic PV (1984) Singular perturbation and time- scale methods in control theory: survey 1976–1983. Automatica 20:273–293
Shimjith SR, Tiwari AP, Bandyopadhyay B (2013) Direct block diagonalization and composite control of three-time-scale systems. In: Modeling and control of a large nuclear reactor. Springer, Berlin Heidelberg, pp 101–114
Shimjith SR, Tiwari AP, Bandyopadhyay B (2011) A three time scale approach for design of linear state regulator for spatial control of advanced heavy water reactor. IEEE Trans Nucl Sci 58:1264–1276
Shouse KR, Taylor DG (1993) Reduced-order discrete-time recursive observers for linear continuous-time singularly perturbed systems. In: Proceedings of the American control conference, pp 2128–2129
Singh NP, Singh YP, Ahson SI (1986) An iterative approach to reduced- order modeling of synchronous machines using singular perturbation. Proc IEEE 74:892–893
Syrcos G, Sannuti P (1983) Singular perturbation modeling of continuous and discrete physical systems. Int J Control 37:1007–1022
Yamamoto K, Yamamoto Y, Nagahara M (2017) Simultaneous rejection of signals below and above the Nyquist frequency. In: IEEE conference on control technology and applications, pp 1135–1139
Yoo H, Gajic Z (2018) New designs of linear observers and observer-based controllers for singularly perturbed linear systems. IEEE Trans Auto Control 63:3904–3911
Yoo H, Gajic Z (2017) New designs of reduced order observer-based controllers for singularly perturbed linear systems. Math Probl Eng 1–14
Zhang Y, Naidu DS, Cai CX, Zou Y (2014) Singular perturbations and time scales in control theories and applications: an overview 2002–2012. Int J Info Syst Sci 9:1–36

# Chapter 2
# Three-Stage Design Using Two-Time-Scale Technique

**Abstract** This chapter introduces a three-stage feedback control design for a three-time-scale system using a two-time-scale approach. Through two-time-scale similarity transformations, the system is converted into a lower triangular form with slow, fast, and very fast subsystems. Feedback controls are then separately designed for these subsystems, employing a two-time-scale methodology. The final feedback control for the full-order system is derived from subsystem controls using state variable transformation matrices. The method is simple and easy to implement. It requires only four linear transformations and minimizes matrix calculations. Also, it significantly reduces computation time compared to regular three-time-scale approaches.

**Keywords** Feedback control · Singular perturbation · Three-stage design · Three-time-scale system · Two-time-scale system

## 2.1 Introduction

The exact mathematical representation of a physical system is a key issue for appropriate control design. Due to the presence of certain small parameters, the realistic description of many physical systems leads to high system order. Singularly perturbed systems are those in which the suppression of a small parameter leads to the degeneration of the system order (Naidu 1988). Designing a feedback control for such a system is challenging due to coupled dynamics with wide speed variations as well as the system's high-order structure. This results in the emergence of the two-time-scale and three-time-scale phenomena. The literature (Kokotovic et al. 1976; Saksena et al. 1984; Zhang et al. 2014) extensively discusses singular perturbation and time-scale methods. While three-time-scale systems have 'slow', 'fast', and 'very fast' modes, two-time-scale systems only have 'slow' and 'fast' dynamic modes. Feedback control for such a multi-time-scale system is constructed in stages that are sequential, such as two stages for systems with two-time-scales and three stages for systems with three-time-scales.

The explicit invertible block-diagonal transformation-based two-stage feedback control architecture for a two-time-scale system is elaborated in Phillips (1980).

 
R. Munje et al., *Multi-stage Feedback Control Design for Multi-time-scale System*, Springer Tracts in Electrical and Electronics Engineering,
https://doi.org/10.1007/978-981-97-4592-0_2

Extending this, the three-stage design for a three-time-scale system is discussed in Naidu and Ravinder (1985). In this, the three-time-scale system is directly block-diagonalized into 'slow', 'fast', and 'very fast' modes using linear transformation matrices and then feedback controls are designed in three independent stages. Further, in Singh et al. (1986), a two-stage decomposition approach is utilized to decouple the three-time-scale system into three lower-order subsystems in the same way as it was done to decouple the two-time-scale system into 'slow' and 'fast' subsystems in Phillips (1980).

Publication by Radisavljevic-Gajic et al. (2017) describes a simplified three-stage feedback control design method for three-time-scale system. In this case, the original three-time-scale system is transformed into an upper triangular form via similarity transformations rather than block-diagonal transformation as done in Naidu and Ravinder (1985). Then, controllers for very fast, fast, and slow subsystems are designed in an order. As a result of the integration of feedback controls, the original system is given a thorough composite state feedback control. The derived feedback control only needs to solve six algebraic equations, as opposed to the standard three-stage approach using block-diagonalization (Naidu and Ravinder 1985), which has to solve nine equations.

This chapter establishes an alternative method of building a three-stage feedback control for a system with slow, fast, and very fast eigenvalues (Munje et al. 2018). First, the original system is transformed, using two similarity transformations of the two-time-scale representation, to the lower triangular form. Then, at three distinct stages, feedback controls are derived for slow, fast, and very fast subsystems. Matrix transformations are used after the first and second stages of building feedback controllers. In the end, a total feedback control is created by combining subsystem feedback controls. In doing so, only two block-triangular transformations and two feedback control design matrix transformations are needed. The complete process of feedback control design is illustrated and explained by a simplified block diagram representation and it is shown that the computational complexity and time, both are reduced to a considerable extent compared to Radisavljevic-Gajic et al. (2017).

## 2.2 Design Methodology

Consider a linear time-invariant controllable continuous-time system as

$$\dot{\mathbf{z}} = \mathbf{A}\mathbf{z} + \mathbf{B}\mathbf{u}, \tag{2.1}$$

$$\mathbf{y} = \mathbf{C}\mathbf{z}, \tag{2.2}$$

where $\mathbf{z} \in \Re^n$, $\mathbf{u} \in \Re^m$ and $\mathbf{y} \in \Re^p$. The matrices $\mathbf{A}$, $\mathbf{B}$ and $\mathbf{C}$ are of appropriate dimensions. Suppose the system (2.1)–(2.2) exhibits three-time-scale property, so that, it can be denoted as

$$\begin{bmatrix} \dot{\mathbf{z}}_1 \\ \dot{\mathbf{z}}_2 \\ \dot{\mathbf{z}}_3 \end{bmatrix} = \begin{bmatrix} \mathbf{A}_{11} & \mathbf{A}_{12} & \mathbf{A}_{13} \\ \mathbf{A}_{21} & \mathbf{A}_{22} & \mathbf{A}_{23} \\ \mathbf{A}_{31} & \mathbf{A}_{32} & \mathbf{A}_{33} \end{bmatrix} \begin{bmatrix} \mathbf{z}_1 \\ \mathbf{z}_2 \\ \mathbf{z}_3 \end{bmatrix} + \begin{bmatrix} \mathbf{B}_1 \\ \mathbf{B}_2 \\ \mathbf{B}_3 \end{bmatrix} \mathbf{u}, \tag{2.3}$$

$$\mathbf{y} = \begin{bmatrix} \mathbf{C}_1 & \mathbf{C}_2 & \mathbf{C}_3 \end{bmatrix} \begin{bmatrix} \mathbf{z}_1^T & \mathbf{z}_2^T & \mathbf{z}_3^T \end{bmatrix}^T, \tag{2.4}$$

where $\mathbf{z}_1 \in \Re^{n_1}, \mathbf{z}_2 \in \Re^{n_2}$ and $\mathbf{z}_3 \in \Re^{n_3}$ such that $(n_1 + n_2 + n_3 = n)$ and the matrices $\mathbf{A}_{ij}$, $\mathbf{B}_i$ and $\mathbf{C}_i$ are of proper orders. Thus, the system (2.3)–(2.4) is having $n_1$ 'slow' modes, $n_2$ 'fast' modes and $n_3$ 'very fast' modes. The multi-time-scale behavior of such a system makes it computationally challenging to construct a regular feedback control. As a result, by regrouping the states of the original three-time-scale system, a feedback controller is constructed in three stages utilizing a two-time-scale technique. The foundation of this design process is similarity transformations. As a result, the assumption that system (2.1) is controllable also means that decoupled subsystems, which were generated by applying similarity transformations at various design stages, are controllable. The existence of inverses of $\mathbf{A}_{22}$ and $\mathbf{A}_{33}$, which is necessary to obtain the solution of algebraic equations, covered in the next subsections, is another fundamental assumption in singular perturbations and time-scale systems (Radisavljevic-Gajic et al. 2017). At first, a block-triangular form is obtained and then the feedback control design is done.

### 2.2.1 *Block-Triangularization*

To simplify the design, let us consider regrouping of states of the system (2.3) as

$$\mathbf{z}_a = \begin{bmatrix} \mathbf{z}_1 \\ \mathbf{z}_2 \end{bmatrix}, \mathbf{z}_b = \mathbf{z}_3. \tag{2.5}$$

Therefore, the system (2.3) can be rewritten as

$$\begin{bmatrix} \dot{\mathbf{z}}_a \\ \dot{\mathbf{z}}_b \end{bmatrix} = \begin{bmatrix} \bar{\mathbf{A}}_{11} & \bar{\mathbf{A}}_{12} \\ \bar{\mathbf{A}}_{21} & \bar{\mathbf{A}}_{22} \end{bmatrix} \begin{bmatrix} \mathbf{z}_a \\ \mathbf{z}_b \end{bmatrix} + \begin{bmatrix} \bar{\mathbf{B}}_1 \\ \bar{\mathbf{B}}_2 \end{bmatrix} \mathbf{u}, \tag{2.6}$$

where $\bar{\mathbf{A}}_{ij}$ and $\bar{\mathbf{B}}_i$ are grouped accordingly. The system (2.6) is the two-time-scale representation of the system (2.3), where $n_1$ 'slow' and $n_2$ 'fast' states are merged to form $(n_1 + n_2)$ 'slow' states and $n_3$ 'very fast' states are taken as 'fast' states. Now, the application of similarity transformation

$$\begin{bmatrix} \mathbf{z}_s \\ \mathbf{z}_b \end{bmatrix} = \begin{bmatrix} \mathbf{I}_{n_1+n_2} & \mathbf{P} \\ \mathbf{0} & \mathbf{I}_{n_3} \end{bmatrix} \begin{bmatrix} \mathbf{z}_a \\ \mathbf{z}_b \end{bmatrix} = \mathbf{T}_1 \begin{bmatrix} \mathbf{z}_a \\ \mathbf{z}_b \end{bmatrix}, \tag{2.7}$$

to the system (2.6), produces

$$\begin{bmatrix} \dot{\mathbf{z}}_s \\ \dot{\mathbf{z}}_b \end{bmatrix} = \begin{bmatrix} \bar{\mathbf{A}}_s & \mathbf{0} \\ \bar{\mathbf{A}}_{21} & \bar{\mathbf{A}}_f \end{bmatrix} \begin{bmatrix} \mathbf{z}_s \\ \mathbf{z}_b \end{bmatrix} + \begin{bmatrix} \bar{\mathbf{B}}_s \\ \bar{\mathbf{B}}_2 \end{bmatrix} \mathbf{u}, \tag{2.8}$$

where $\mathbf{I}_n$ is identity matrix of dimension $n$, $\bar{\mathbf{A}}_s = \bar{\mathbf{A}}_{11} + \mathbf{P}\bar{\mathbf{A}}_{21}$, $\bar{\mathbf{A}}_f = \bar{\mathbf{A}}_{22} - \bar{\mathbf{A}}_{21}\mathbf{P}$ and $\bar{\mathbf{B}}_s = \bar{\mathbf{B}}_1 + \mathbf{P}\bar{\mathbf{B}}_2$. As $\mathbf{A}_{33}$ is assumed to be invertible, the unique solution of $\mathbf{P}$ can be found by solving

$$\bar{\mathbf{A}}_{12} + \mathbf{P}\bar{\mathbf{A}}_{22} - (\bar{\mathbf{A}}_{11} + \mathbf{P}\bar{\mathbf{A}}_{21})\mathbf{P} = 0, \tag{2.9}$$

iteratively. Consequently, the system (2.8) is transformed into lower triangular form, where subsystem $\mathbf{z}_s$ is completely decoupled from $\mathbf{z}_b$ and it is given by

$$\dot{\mathbf{z}}_s = \bar{\mathbf{A}}_s \mathbf{z}_s + \bar{\mathbf{B}}_s \mathbf{u}. \tag{2.10}$$

Since, $\mathbf{z}_s$ combines $n_1$ 'slow' and $n_2$ 'fast' states of the system (2.3), it can also be written in two-time-scale form, by suitable partitioning of $\bar{\mathbf{A}}_s$ and $\bar{\mathbf{B}}_s$, as

$$\begin{bmatrix} \dot{\mathbf{z}}_{s1} \\ \dot{\mathbf{z}}_{s2} \end{bmatrix} = \begin{bmatrix} \bar{\mathbf{A}}_{s11} & \bar{\mathbf{A}}_{s12} \\ \bar{\mathbf{A}}_{s21} & \bar{\mathbf{A}}_{s22} \end{bmatrix} \begin{bmatrix} \mathbf{z}_{s1} \\ \mathbf{z}_{s2} \end{bmatrix} + \begin{bmatrix} \bar{\mathbf{B}}_{s1} \\ \bar{\mathbf{B}}_{s2} \end{bmatrix} \mathbf{u}, \tag{2.11}$$

where $\mathbf{z}_s = \begin{bmatrix} \mathbf{z}_{s1}^T & \mathbf{z}_{s2}^T \end{bmatrix}^T$ with $\mathbf{z}_{s1} \in \Re^{n_1}$ and $\mathbf{z}_{s2} \in \Re^{n_2}$. At this moment, a linear transformation

$$\begin{bmatrix} \mathbf{z}_{ss} \\ \mathbf{z}_{s2} \end{bmatrix} = \begin{bmatrix} \mathbf{I}_{n_1} & \mathbf{P}_s \\ \mathbf{0} & \mathbf{I}_{n_2} \end{bmatrix} \begin{bmatrix} \mathbf{z}_{s1} \\ \mathbf{z}_{s2} \end{bmatrix} = \mathbf{T}_2 \begin{bmatrix} \mathbf{z}_{s1} \\ \mathbf{z}_{s2} \end{bmatrix}, \tag{2.12}$$

is applied to the system (2.11). Here, matrix $\mathbf{P}_s$ is determined by evaluating

$$\bar{\mathbf{A}}_{s12} + \mathbf{P}_s\bar{\mathbf{A}}_{s22} - (\bar{\mathbf{A}}_{s11} + \mathbf{P}_s\bar{\mathbf{A}}_{s21})\mathbf{P}_s = 0. \tag{2.13}$$

Solution of $\mathbf{P}_s$ can be reached under the assumption that $\mathbf{A}_{22}$ is invertible. Accordingly, a lower triangular form of (2.11) is acquired as

$$\begin{bmatrix} \dot{\mathbf{z}}_{ss} \\ \dot{\mathbf{z}}_{s2} \end{bmatrix} = \begin{bmatrix} \bar{\mathbf{A}}_{ss} & \mathbf{0} \\ \bar{\mathbf{A}}_{s21} & \bar{\mathbf{A}}_{sf} \end{bmatrix} \begin{bmatrix} \mathbf{z}_{s1} \\ \mathbf{z}_{s2} \end{bmatrix} + \begin{bmatrix} \bar{\mathbf{B}}_{ss} \\ \bar{\mathbf{B}}_{s2} \end{bmatrix} \mathbf{u}, \tag{2.14}$$

where $\bar{\mathbf{A}}_{ss} = \bar{\mathbf{A}}_{s11} + \mathbf{P}_s\bar{\mathbf{A}}_{s21}$, $\bar{\mathbf{A}}_{sf} = \bar{\mathbf{A}}_{s22} - \bar{\mathbf{A}}_{s21}\mathbf{P}_s$ and $\bar{\mathbf{B}}_{ss} = \bar{\mathbf{B}}_{s1} + \mathbf{P_s}\bar{\mathbf{B}}_{s2}$. Thus, using block-triangular forms (2.8) and (2.14) of two-time-scale structures, the original three-time-scale system (2.3) is changed into block-triangular form as

$$\begin{bmatrix} \dot{\mathbf{z}}_{ss} \\ \dot{\mathbf{z}}_{s2} \\ \dot{\mathbf{z}}_3 \end{bmatrix} = \begin{bmatrix} \bar{\mathbf{A}}_{ss} & \mathbf{0} & \mathbf{0} \\ \bar{\mathbf{A}}_{s21} & \bar{\mathbf{A}}_{sf} & \mathbf{0} \\ \bar{\mathbf{A}}_{31} & \bar{\mathbf{A}}_{32} & \bar{\mathbf{A}}_f \end{bmatrix} \begin{bmatrix} \mathbf{z}_{ss} \\ \mathbf{z}_{s2} \\ \mathbf{z}_3 \end{bmatrix} + \begin{bmatrix} \bar{\mathbf{B}}_{ss} \\ \bar{\mathbf{B}}_{s2} \\ \bar{\mathbf{B}}_2 \end{bmatrix} \mathbf{u}, \tag{2.15}$$

where $\bar{\mathbf{A}}_{21}$ is suitably partitioned as $[\bar{\mathbf{A}}_{31} \quad \bar{\mathbf{A}}_{32}]$. Now, one can continue for a three-stage feedback control design starting with the 'slow' subsystem $\mathbf{z}_{ss}$.

### 2.2.2 *Three-Stage Feedback Control Design*

The three-stage feedback control design is described using Theorems 2.1 and 2.2.

**Theorem 2.1** *If $\mathbf{K}_{ss}$ and $\mathbf{K}_{sf}$ indicate respectively the stabilizing feedback gains for the slow and fast subsystems in (2.14), i.e.,*

$$\psi_i(\bar{\mathbf{A}}_{ss} - \bar{\mathbf{B}}_{ss}\mathbf{K}_{ss}) < 0 \qquad \textit{for} \quad i = 1, \ldots, n_1,$$
$$\psi_j(\bar{\mathbf{A}}_{sf} - \bar{\mathbf{B}}_{sf}\mathbf{K}_{sf}) < 0 \qquad \textit{for} \quad j = 1, \ldots, n_2,$$

*such that, $max(\psi_i) \ll min(\psi_j)$, where $\bar{\mathbf{B}}_{sf} = \bar{\mathbf{B}}_{s2} - \mathbf{M}_s\bar{\mathbf{B}}_{ss}$ in which $\mathbf{M}_s$ satisfies*

$$\bar{\mathbf{A}}_{s21} - \bar{\mathbf{B}}_{s2}\mathbf{K}_{ss} - \mathbf{M}_s(\bar{\mathbf{A}}_{ss} - \bar{\mathbf{B}}_{ss}\mathbf{K}_{ss}) + \bar{\mathbf{A}}_{sf}\mathbf{M}_s = 0. \tag{2.16}$$

*Then, there exists a $\mathbf{K}_s$ that stabilizes (2.11), given by*

$$\mathbf{K}_s = \left[\mathbf{K}_{ss} - \mathbf{K}_{sf}\mathbf{M}_s \;\; (\mathbf{K}_{ss} - \mathbf{K}_{sf}\mathbf{M}_s)\mathbf{P}_s + \mathbf{K}_{sf}\right]. \tag{2.17}$$

***Proof*** In the first stage, applying the feedback control

$$\mathbf{u}_s = -\mathbf{K}_{ss}\mathbf{z}_{ss} + \mathbf{u}_{sf}, \tag{2.18}$$

to the system (2.14), leads to

$$\begin{bmatrix} \dot{\mathbf{z}}_{ss} \\ \dot{\mathbf{z}}_{s2} \end{bmatrix} = \begin{bmatrix} \bar{\mathbf{A}}_{ss} - \bar{\mathbf{B}}_{ss}\mathbf{K}_{ss} & \mathbf{0} \\ \bar{\mathbf{A}}_{s21} - \bar{\mathbf{B}}_{s2}\mathbf{K}_{ss} & \bar{\mathbf{A}}_{sf} \end{bmatrix} \begin{bmatrix} \mathbf{z}_{ss} \\ \mathbf{z}_{s2} \end{bmatrix} + \begin{bmatrix} \bar{\mathbf{B}}_{ss} \\ \bar{\mathbf{B}}_{s2} \end{bmatrix} \mathbf{u}_{sf}. \tag{2.19}$$

Now, using the change of the variables as

$$\begin{bmatrix} \mathbf{z}_{ss} \\ \mathbf{z}_{sf} \end{bmatrix} = \begin{bmatrix} \mathbf{I}_{n_1} & \mathbf{0} \\ -\mathbf{M}_s & \mathbf{I}_{n_2} \end{bmatrix} \begin{bmatrix} \mathbf{z}_{ss} \\ \mathbf{z}_{s2} \end{bmatrix} = \mathbf{T}_3 \begin{bmatrix} \mathbf{z}_{ss} \\ \mathbf{z}_{s2} \end{bmatrix}, \tag{2.20}$$

the system (2.19) is revised to block-diagonal form

$$\begin{bmatrix} \dot{\mathbf{z}}_{ss} \\ \dot{\mathbf{z}}_{sf} \end{bmatrix} = \begin{bmatrix} \bar{\mathbf{A}}_{ss} - \bar{\mathbf{B}}_{ss}\mathbf{K}_{ss} & \mathbf{0} \\ \mathbf{0} & \bar{\mathbf{A}}_{sf} \end{bmatrix} \begin{bmatrix} \mathbf{z}_{ss} \\ \mathbf{z}_{sf} \end{bmatrix} + \begin{bmatrix} \bar{\mathbf{B}}_{ss} \\ \bar{\mathbf{B}}_{sf} \end{bmatrix} \mathbf{u}_{sf}. \tag{2.21}$$

In (2.21), subsystem $\mathbf{z}_{sf}$ is made totally independent of $\mathbf{z}_{ss}$. Therefore, in the second stage

$$\mathbf{u}_{sf} = -\mathbf{K}_{sf}\mathbf{z}_{sf}, \tag{2.22}$$

is applied to the system (2.21) as

$$\begin{bmatrix} \dot{\mathbf{z}}_{ss} \\ \dot{\mathbf{z}}_{sf} \end{bmatrix} = \begin{bmatrix} \bar{\mathbf{A}}_{ss} - \bar{\mathbf{B}}_{ss}\mathbf{K}_{ss} & -\bar{\mathbf{B}}_{ss}\mathbf{K}_{sf} \\ \mathbf{0} & \bar{\mathbf{A}}_{sf} - \bar{\mathbf{B}}_{sf}\mathbf{K}_{sf} \end{bmatrix} \begin{bmatrix} \mathbf{z}_{ss} \\ \mathbf{z}_{sf} \end{bmatrix}. \tag{2.23}$$

Thus, the overall control input for stabilizing the system (2.11) is computed using transformations (2.12) and (2.20) and controls (2.18) and (2.22) as

$$\mathbf{u}_s = -\mathbf{K}_s\mathbf{z}_s, \tag{2.24}$$

where

$$\mathbf{K}_s = \left[\mathbf{K}_{ss} - \mathbf{K}_{sf}\mathbf{M}_s \ (\mathbf{K}_{ss} - \mathbf{K}_{sf}\mathbf{M}_s)\mathbf{P}_s + \mathbf{K}_{sf}\right]. \tag{2.25}$$

□

Recall that, the controllability of (2.1) confirms the controllability properties of $(\bar{\mathbf{A}}_{ss}, \bar{\mathbf{B}}_{ss})$ and $(\bar{\mathbf{A}}_{sf}, \bar{\mathbf{B}}_{sf})$. Also, it must be noted that the Eq. (2.16) has exact solution if the matrices $(\bar{\mathbf{A}}_{ss} - \bar{\mathbf{B}}_{ss}\mathbf{K}_{ss})$ and $\bar{\mathbf{A}}_{sf}$ do not have common eigenvalues. And, this is satisfied because $(\bar{\mathbf{A}}_{ss} - \bar{\mathbf{B}}_{ss}\mathbf{K}_{ss})$ is user designed with desired 'slow' eigenvalues, whereas $\bar{\mathbf{A}}_{sf}$ has original system's 'fast' eigenvalues.

**Theorem 2.2** *If the stabilizing feedback gains, for the slow and fast subsystems in (2.8), are respectively given by* $\mathbf{K}_s$ *(designed in Theorem 2.1) and* $\mathbf{K}_f$*, i.e.,*

$$\psi_k(\bar{\mathbf{A}}_s - \bar{\mathbf{B}}_s\mathbf{K}_s) < 0 \quad \text{for} \quad k = 1, \ldots, n_1 + n_2,$$
$$\psi_l(\bar{\mathbf{A}}_f - \bar{\mathbf{B}}_f\mathbf{K}_f) < 0 \quad \text{for} \quad l = 1, \ldots, n_3,$$

*such that,* $max(\psi_k) \ll min(\psi_l)$*, where* $\bar{\mathbf{B}}_f = \bar{\mathbf{B}}_2 - \mathbf{M}\bar{\mathbf{B}}_s$ *with* $\mathbf{M}$ *satisfying*

$$\bar{\mathbf{A}}_{21} - \bar{\mathbf{B}}_2\mathbf{K}_s - \mathbf{M}(\bar{\mathbf{A}}_s - \bar{\mathbf{B}}_s\mathbf{K}_s) + \bar{\mathbf{A}}_f\mathbf{M}_s = 0. \tag{2.26}$$

*Then, the existence of* $\mathbf{K}$*, given below is ensured*

$$\mathbf{K} = \left[\mathbf{K}_s - \mathbf{K}_f\mathbf{M} \ (\mathbf{K}_s - \mathbf{K}_f\mathbf{M})\mathbf{P} + \mathbf{K}_f\right] \tag{2.27}$$

*which stabilizes the system (2.6) and thereby (2.1).*

***Proof*** As the control (2.24) is for (2.10), which is the decoupled part of (2.8), it can also be written in the form

$$\mathbf{u} = -\mathbf{K}_s\mathbf{z}_s + \mathbf{u}_f \tag{2.28}$$

and given to the system (2.8). This results in

$$\begin{bmatrix} \dot{\mathbf{z}}_s \\ \dot{\mathbf{z}}_b \end{bmatrix} = \begin{bmatrix} \bar{\mathbf{A}}_s - \bar{\mathbf{B}}_s\mathbf{K}_s & \mathbf{0} \\ \bar{\mathbf{A}}_{21} - \bar{\mathbf{B}}_2\mathbf{K}_s & \bar{\mathbf{A}}_f \end{bmatrix} \begin{bmatrix} \mathbf{z}_s \\ \mathbf{z}_b \end{bmatrix} + \begin{bmatrix} \bar{\mathbf{B}}_s \\ \bar{\mathbf{B}}_2 \end{bmatrix} \mathbf{u}_f. \tag{2.29}$$

Once again introducing a change of coordinates as

$$\begin{bmatrix} \mathbf{z}_s \\ \mathbf{z}_f \end{bmatrix} = \begin{bmatrix} \mathbf{I}_{n_1+n_2} & \mathbf{0} \\ -\mathbf{M} & \mathbf{I}_{n_3} \end{bmatrix} \begin{bmatrix} \mathbf{z}_s \\ \mathbf{z}_b \end{bmatrix} = \mathbf{T}_4 \begin{bmatrix} \mathbf{z}_s \\ \mathbf{z}_b \end{bmatrix}, \tag{2.30}$$

for the system (2.29), subsystem $\mathbf{z}_f$ is isolated from $\mathbf{z}_s$ as

$$\begin{bmatrix} \dot{\mathbf{z}}_s \\ \dot{\mathbf{z}}_f \end{bmatrix} = \begin{bmatrix} \bar{\mathbf{A}}_s - \bar{\mathbf{B}}_s\mathbf{K}_s & \mathbf{0} \\ \mathbf{0} & \bar{\mathbf{A}}_f \end{bmatrix} \begin{bmatrix} \mathbf{z}_s \\ \mathbf{z}_f \end{bmatrix} + \begin{bmatrix} \bar{\mathbf{B}}_s \\ \bar{\mathbf{B}}_f \end{bmatrix} \mathbf{u}_f. \tag{2.31}$$

Finally, in the third stage

$$\mathbf{u}_f = -\mathbf{K}_f\mathbf{z}_f, \tag{2.32}$$

is applied to the system (2.31) as

$$\begin{bmatrix} \dot{\mathbf{z}}_s \\ \dot{\mathbf{z}}_f \end{bmatrix} = \begin{bmatrix} \bar{\mathbf{A}}_s - \bar{\mathbf{B}}_s\mathbf{K}_s & -\bar{\mathbf{B}}_s\mathbf{K}_f \\ \mathbf{0} & \bar{\mathbf{A}}_f - \bar{\mathbf{B}}_f\mathbf{K}_f \end{bmatrix} \begin{bmatrix} \mathbf{z}_s \\ \mathbf{z}_f \end{bmatrix}. \tag{2.33}$$

Therefore, the overall control for the system (2.6), consequently for system (2.1), can be obtained from (2.7) and (2.30) and controls (2.28) and (2.32) as

$$\mathbf{u} = -\mathbf{K}\mathbf{z}, \tag{2.34}$$

where

$$\mathbf{K} = \begin{bmatrix} \mathbf{K}_s - \mathbf{K}_f\mathbf{M} & (\mathbf{K}_s - \mathbf{K}_f\mathbf{M})\mathbf{P} + \mathbf{K}_f \end{bmatrix}. \tag{2.35}$$

□

Again, the controllability of $(\bar{\mathbf{A}}_f, \bar{\mathbf{B}}_f)$ is preserved due to controllability of (2.1). Further, the assumption needed to solve (2.26) is that there are no common eigenvalues between $(\bar{\mathbf{A}}_s - \bar{\mathbf{B}}_s\mathbf{K}_s)$ and $\bar{\mathbf{A}}_f$, which can be fulfilled by the fact that $(\bar{\mathbf{A}}_s - \bar{\mathbf{B}}_s\mathbf{K}_s)$ is designed with user-chosen eigenvalues laying in slow and fast dynamic modes and $\bar{\mathbf{A}}_f$ has very fast eigenvalues of full-order system. Figure 2.1 shows a MATLAB/Simulink (2019) block schematic of the entire three-stage feedback control design process, complete with appropriate dimensions for the transformation and feedback gain matrices. Matrix transformations are depicted in blue in Fig. 2.1 while the steps of feedback control designs are shown in red. To obtain the block-triangular form of the original system (2.3), similarity transformations (2.7) and (2.12) are

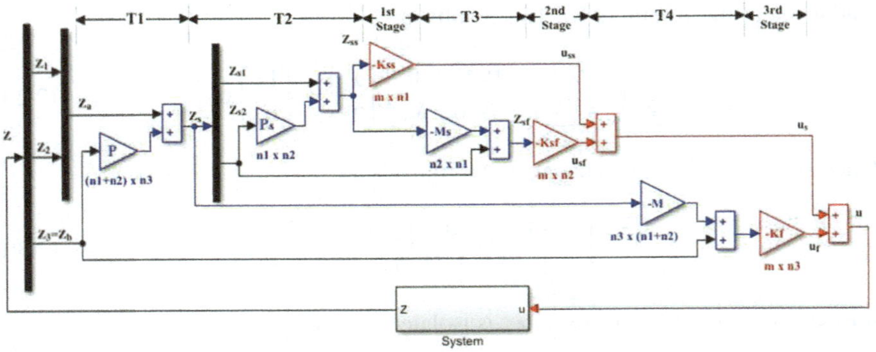


**Fig. 2.1** Representation of the three-stage feedback control design in Simulink

employed. Although the dimensions of these two transformations are different, their structures are identical. The next two transformations (2.20) and (2.30), which have identical representations but different orders. These are used to convert a block-triangular system into a block-diagonal form. It is crucial to keep in mind that the lower triangular form (2.15) of system (2.3) can only be obtained by solving two nonsymmetric Riccati algebraic equations (2.9) and (2.13) for $\mathbf{P}$ and $\mathbf{P}_s$ respectively and a three-stage is performed by solving two Sylvester equations (2.16) and (2.26) separately for $\mathbf{M}_s$ and $\mathbf{M}$, i.e., only four algebraic equations are required to be solved compared to six equations in Radisavljevic-Gajic et al. (2017). Further, because computations are performed using a two-time-scale representation rather than a three-time-scale representation, there is a significant reduction in processing requirements without sacrificing accuracy or the ability to construct various local feedback controllers for local subsystems.

**Remark 2.1** If the 'very fast' eigenvalues of the original system are supposed to be stable, then just the first two steps need to be finished for design purposes. The first three algebraic equations can therefore be solved to obtain an approximate feedback control $\bar{\mathbf{K}}$ of the entire system. It can be realized by setting $\mathbf{K}_f = 0$ in (2.35) as

$$\bar{\mathbf{K}} = \left[\mathbf{K}_s \ \mathbf{K}_s\mathbf{P}\right]. \tag{2.36}$$

where $\mathbf{K}_s$ is given by (2.25).

**Remark 2.2** If the controller is employed to stabilize slow eigenvalues alone and 'fast' and 'very fast' eigenvalues are asymptotically stable, then the solution of just two algebraic equations can result in a stabilizing reduced-order feedback gain $\tilde{\mathbf{K}}$. This can be obtained by solving (2.25) and (2.35) for $\mathbf{K}_{sf} = 0$ and $\mathbf{K}_f = 0$ respectively. The approximation to overall feedback gain will be

$$\tilde{\mathbf{K}} = \left[\bar{\mathbf{K}}_s \ \bar{\mathbf{K}}_s\mathbf{P}\right]. \tag{2.37}$$

where $\bar{\mathbf{K}}_s = [\mathbf{K}_{ss} \quad \mathbf{K}_{ss}\mathbf{P}_s]$.

## 2.3 Numerical Example

The efficacy of the devised control strategy is demonstrated by solving a numerical example and comparing the execution time with the control algorithm investigated in Radisavljevic-Gajic et al. (2017). Consider the ninth-order model (Naidu and Ravinder 1985) with state and input matrices as

$$\mathbf{A} = \begin{bmatrix} -0.31 & 0 & 1.03 & 0 & 0 & 2.00 & 0.10 & -6.30 & -0.40 \\ 1.01 & -0.65 & -0.08 & 0 & 0.83 & 1.70 & 3.00 & 2.40 & 0.20 \\ -0.11 & 0 & -1.26 & 0 & -6.30 & -1.04 & 0 & 4.10 & 0.01 \\ 0 & 0.10 & 0 & -13.00 & 0 & 0.48 & 0 & 0 & 0 \\ 0 & -0.05 & 0.25 & 0 & -17.60 & -0.30 & 0 & 0 & 0 \\ 0 & 4.50 & -3.80 & -640.00 & 375.00 & -97.50 & 1.12 & 0 & -0.80 \\ 0 & -2.9000 & 2.75 & 11.10 & 0 & 1.01 & -143.00 & 0.50 & 0 \\ 2.50 & -5.70 & -4.00 & 0 & 0 & -0.91 & 0 & -184.00 & -4.20 \\ 1.80 & -3.10 & 3.18 & 0 & 1.76 & -0.70 & 1.53 & 3.20 & -207.00 \end{bmatrix},$$

$$\mathbf{B} = \begin{bmatrix} 0.8812 & 1.1366 & -0.6790 & -0.1311 & -0.2270 & 0.9823 & -1.0174 & -0.8227 & 0.7900 \end{bmatrix}^T.$$

State matrix **A** has three distinct modes of eigenvalues, namely, slow mode (−1.5627, −0.8018, −0.3046), fast mode (−20.3805, −15.3097) and very fast mode (−206.4022, −184.3452, −142.9591, −92.2553). It is verified that the system is controllable. For this system, feedback control is derived by the three-stage design methodology explained in Sect. 2.2. The above system is converted into block-triangular form by iteratively evaluating (2.9) and (2.13) for $\mathbf{P}$ and $\mathbf{P}_s$ respectively. After 500 iterations corresponding values of $\mathbf{P}$ and $\mathbf{P}_s$ are

$$\mathbf{P} = \begin{bmatrix} 0.0222 & 0.0008 & -0.0345 & -0.0013 \\ 0.0184 & 0.0212 & 0.0134 & 0.0006 \\ -0.0119 & -0.0001 & 0.0224 & -0.0004 \\ 0.0060 & 0.0000 & -0.0000 & -0.0000 \\ -0.0040 & -0.0000 & -0.0000 & 0.0000 \end{bmatrix}, \mathbf{P}_s = \begin{bmatrix} -0.8283 & 0.3787 \\ -0.6139 & 0.3246 \\ 0.3960 & -0.5569 \end{bmatrix}.$$

Here, it is confirmed that the eigenvalues of $\bar{\mathbf{A}}_{ss}$, $\bar{\mathbf{A}}_{sf}$, $\bar{\mathbf{A}}_f$ are the same as that of the eigenvalues of slow, fast and very fast modes respectively. In the first stage, feedback gain $\mathbf{K}_{ss}$ is designed to place eigenvalues of $(\bar{\mathbf{A}}_{ss} - \bar{\mathbf{B}}_{ss}\mathbf{K}_{ss})$ at desired location of (−0.34, −0.58, −0.78) using 'place' command (Matlab 2019). The feedback gain is calculated as

$$\mathbf{K}_{ss} = \begin{bmatrix} -0.1347 & 0.2549 & 1.7898 \end{bmatrix}.$$

Then, the system is again changed to block-diagonal form by determining $\mathbf{M}_s$ after 500 iterations as

$$\mathbf{M}_s = \begin{bmatrix} -0.0018 & 0.0102 & 0.0187 \\ -0.0017 & 0.0007 & 0.0389 \end{bmatrix}.$$

In the second stage, eigenvalues of the fast modes are to be placed at $(-12, -17)$ using feedback gain $\mathbf{K}_{sf}$. This gain is estimated as

$$\mathbf{K}_{sf} = \left[-25.4497\ 47.8020\right].$$

At the end of second stage, feedback gain $\mathbf{K}_s$, as given in (2.25), is formulated as

$$\mathbf{K}_s = \left[-0.1019\ 0.4794\ 0.4039\ -25.4996\ 47.6940\right].$$

For designing the feedback gain in the third stage, the system is block-diagonalized by solving (2.30) iteratively once again 500 times to get

$$\mathbf{M} = \begin{bmatrix} 0.0002 & 0.0592 & -0.0617 & -8.3201 & 4.7670 \\ -0.0005 & -0.0166 & 0.0207 & -0.1758 & 0.3983 \\ 0.0131 & -0.0289 & -0.0206 & -0.0804 & 0.2054 \\ 0.0094 & -0.0178 & 0.0142 & 0.1316 & -0.1988 \end{bmatrix}.$$

With this, feedback gain $\mathbf{K}_f$ is determined to place eigenvalues of very fast mode at $(-115 - 162 - 190 - 210)$ as

$$\mathbf{K}_f = \left[38.0691\ -12.6643\ -3.2850\ 1.9019\right].$$

The overall feedback gain as per (2.35) for the original system is obtained as

$$\mathbf{K} = \left[-0.0915\ -2.0452\ 2.9202\ 288.4993\right.$$
$$\left.-127.6825\ 40.2411\ -12.6956\ -3.2418\ 1.8898\right].$$

With the application of this composite gain to the original system, the eigenvalues of all the modes of the system are moved to the desired locations with 100% accuracy.

It is important to note that the value of composite gain is identical to that determined by the method described in Radisavljevic-Gajic et al. (2017). But when using the suggested approach, just four algebraic equations rather than six are solved. The computation time comparison is presented in Table 2.1 when this example is simulated using the method examined in Radisavljevic-Gajic et al. (2017) with an equal number of iterations. The percentage saving in the computational time is seen to be 27.53 ms. The amount of computing time depends on the order of the system, the orders of its subsystems, the ratios of slow to fast to very fast eigenvalue modes, and the position of the desired poles.

**Table 2.1** Comparison of computation time

| Computation time (ms) | | Percentage |
|---|---|---|
| Method of Radisavljevic-Gajic et al. (2017) | Proposed | Saving |
| 77.81 | 56.39 | 27.53 |

## 2.4 Conclusion

This chapter presents a three-stage feedback control scheme from the perspective of improving computation time. The feedback control for a three-time-scale system is designed using just four similarity transformations. Even though two-time-scale representations were used to complete the design, the option to design independent local feedback controllers was not lost. Further, there is a significant reduction in design complexity. This control technique may be used to develop controllers for large-scale systems where computational efforts are of utmost significance.

## References

Kokotovic PV, O'Malley RE, Sannuti P (1976) Singular perturbation and order reduction in control theory-an overview. Automatica 12:123–132

MATLAB/Simulink Control Design Toolbox User Manual, MathWorks, Version 4.5 (2019)

Munje RK, Gu D, Xu B, Zhang W (2018) Feedback control design for three-time-scale system using two-time-scale approach. In: Proceedings Chinese control conference, Wuhan, pp 4982–4987

Naidu DS, Ravinder R (1985) On three time scale analysis. In: Proceeding IEEE conference on decision and control, pp 81–85

Naidu DS (1988) Singular perturbation methodology in control systems. Peter Peregrinus Ltd, London U.K

Phillips R (1980) A two-stage design of linear feedback controls. IEEE Trans Auto Control 25:1220–1223

Radisavljevic-Gajic V, Milanovic M, Clayton G (2017) Three-stage feedback controller design with application to three-time-scale linear control system. ASME J Dyn Syst, Meas, Control 139:1–10

Saksena VR, O'Reilly J, Kokotovic PV (1984) Singular perturbation and time- scale methods in control theory: survey 1976–1983. Automatica 20:273–293

Singh NP, Singh YP, Ahson SI (1986) An iterative approach to reduced- order modeling of synchronous machines using singular perturbation. Proc IEEE 74:892–893

Zhang Y, Naidu DS, Cai CX, Zou Y (2014) Singular perturbations and time scales in control theories and applications: an overview 2002–2012. Int J Info Syst Sci 9:1–36

# Chapter 3
# Multi-stage Observer Designs for Three-Time-Scale System with Application to AHWR

**Abstract** The chapter focuses on the ninetieth-order system of the Advanced Heavy Water Reactor (AHWR), characterized by a complex three-time-scale structure. The system comprises 38 slow, 35 fast, and 17 very fast state variables, necessitating an effective controller for spatial power regulation. The study proposes full and reduced-order observers using a novel method for this intricate three-time-scale system. The key innovation lies in multi-stage observer designs, streamlining complexity and computational time while maintaining the flexibility to independently design observer gains for the subsystems. These observers are then applied to estimate the AHWR state, facilitating spatial stabilization through feedback control. Performance evaluations, conducted on the nonlinear AHWR model amidst transients, compare these observers with a controller (without an observer) and a fast output sampling-based controller.

**Keywords** Advanced heavy water reactor · State observer · State estimation · Three-time-scale system · Two-time-scale system

## 3.1 Introduction

Global acceptance exists for nuclear energy as a practical, cost-effective, and ecologically benign way to reduce greenhouse gas emissions from the burning of fossil fuels in the near future. As a result, it is predicted that global nuclear energy production will rise to 554 GWe by 2030 and, to 874 GWe by 2050 (Energy 2017). This brings up concerns about nuclear power plant's stability and safety. In this sense, creating effective power control in a nuclear reactor requires precise knowledge of the immeasurable state variables. For several decades, the measurement of nuclear power plant state variables has been studied (Li et al. 2016). In Park and Cho (1992), a Luenberger-type observer uses the output measurements to estimate xenon, iodine, and neutron flux. However, a robust observer is required for the state estimate of nuclear reactors since the dynamics of these reactors depend on fuel burn-up, spatial power distribution, and the creation of xenon and delayed neutrons. As a result, Wang et al. (2001) and Esteki et al. (2015) submitted proposals for robust sliding mode

 
R. Munje et al., *Multi-stage Feedback Control Design for Multi-time-scale System*, Springer Tracts in Electrical and Electronics Engineering,
https://doi.org/10.1007/978-981-97-4592-0_3

observers (SMOs). However, the system's performance degrades as a result of the unwanted chattering that goes along with the SMO. Therefore, it is recommended to use higher order SMOs (HOSMOs) to monitor the amounts of precursors in Hussain et al. (2013a), fuel temperature and reactivity (Hussain et al. 2013b), and concentrations of xenon and samarium (Yeganeh and Ansarifar 2018). Even though HOSMOs provide nearness between the computed and actual state variables, higher values of control input and tracking errors of HOSMOs make them less suited for real-time implementation.

Accurate mathematical modeling of reactivity is difficult as it relies on several elements. Hence, reactivity is considered as an extended state variable to determine neutron flux and precursor concentrations (Dong et al. 2016). Again in Patel et al. (2018), a multi-scale extended Kalman filtering method based on wavelet transform is suggested. Every estimation algorithm that was previously suggested is tested using a nuclear reactor's point kinetic model (PKM) that has a variable number of delayed neutron groups. However, PKM does not apply to large-sized nuclear reactors because of the significant fluctuation in the neutron flux shape. As a result, a reduced-order linear observer is investigated on the fifty-sixth-order model of the pressurized heavy water reactor for the sole purpose of taking the flux shape variation into account (Tiwari and Bandyopadhyay 1998). It is important to note that large-sized nuclear reactors fall into a unique category of systems known as singularly perturbed systems. Ill-conditioning issues can arise in the direct construction of a full-order or even reduced-order observer for such systems. The optimal option, in this case, is an observer created via time-scale separation, where the primary system is divided into lower-order subsystems with well-conditioned matrices (Zhang et al. 2014). Consequently, it is possible to complete independent designs for the subsystem observers and then combine the outcomes to provide a composite observer gain for the main system. Full- and reduced-order observer designs (O'Reilly 1980; Shouse and Taylor 1993) for the singularly perturbed two-time-scale system formulated by quasi-steady-state modeling involves approximation. Yoo and Gajic (2017, 2018) have recently used direct block-diagonalization of the singularly perturbed two-time-scale system to develop new designs for full- and reduced-order observers with good accuracy.

As the stochastic estimators (Patel et al. 2018) are constructed using a probability distribution function, they are incredibly flexible and dependable for rapid state estimation. However, they are impractical for singularly perturbed multi-time-scale systems such as two- and three-time-scale systems due to their computational complexity. In addition, the occurrence of several eigenvalue clusters with significantly varied speeds causes the three-time-scale system to have much more numerical stiffness than the two-time-scale system. Besides the nonlinear observers (Hussain et al. 2013a, b; Yeganeh and Ansarifar 2018) function flawlessly for the nonlinear plants, their applicability in real-world scenarios is restricted. Conversely, because of their important roles in theory and practice, the use of linear observers is highly valued (Tsui 2015; Yang et al. 2016). This chapter proposes linear observers for the singularly perturbed three-time-scale system, inspired by Yoo and Gajic (2017, 2018). The current work is motivated in part by the dearth of research on the spatial control of advanced heavy water reactors (AHWR) through observers, in addition to the wide range of feedback control suggestions that have been proposed for AHWRs (Munje et al. 2016).

This chapter proposes, respectively, three-stage and two-stage designs for full- and reduced-order observers for the singularly perturbed three-time-scale system (Munje et al. 2018). While the reduced-order observer computes only some states, the full-order observer computes all states. Unlike (Radisavljevic-Gajic et al. 2017), which used a three-time-scale transformation, the three-stage design is implemented utilizing two-time-scale similarity transformations. Additionally, compared to Yoo and Gajic (2018) and Radisavljevic-Gajic et al. (2017), respectively, fewer algebraic equations need to be solved to complete the two- and three-stage designs. Consequently, there is a large reduction in both the design complexity and computing time. Observers created using the described approach are then integrated with the feedback controller, designed in Munje et al. (2015). By comparing the results with the controller without an observer and fast output sampling-based controller, the performance of the nonlinear model of the AHWR is examined under transients.

## 3.2 AHWR Plant Description

The AHWR core is split into 17 fictional nodes that are relatively large for mathematical modeling purposes. The complete model is delineated by following nonlinear equations (Shimjith et al. 2011).

$$\frac{dP_i}{dt} = (\rho_i - \alpha_{ii} - \beta)\frac{P_i}{\ell} + \sum_{j=1}^{17}\alpha_{ji}\frac{P_j}{\ell} + \lambda C_i, \tag{3.1}$$

$$\frac{dC_i}{dt} = \frac{\beta}{\ell}P_i - \lambda C_i, \tag{3.2}$$

$$\frac{dI_i}{dt} = \gamma_I \Sigma_{fi} P_i - \lambda_I I_i, \tag{3.3}$$

$$\frac{dX_i}{dt} = \gamma_X \Sigma_{fi} P_i + \lambda_I I_i - (\lambda_X + \bar{\sigma}_{Xi} P_i)X_i, \tag{3.4}$$

$$\frac{dH_k}{dt} = \kappa v_k, \tag{3.5}$$

$$e_{vx_i}\frac{dx_i}{dt} = P_i - q_{d_i}(h_w - h_d) - q_{d_i}x_i h_c, \tag{3.6}$$

$$e_{x_i}\frac{dh_d}{dt} = q_f(\hat{k}_2 h_f - \hat{k}_1) - q_d(\hat{k}_2 h_d - \hat{k}_1), \tag{3.7}$$

where $i = 1, 2, \ldots, 17$ and $k = 2, 4, 6, 8$. $P$, $C$, $I$, $X$, $H$, $x$ and $h_d$ are respectively nodal powers, single group delayed neutron precursor, iodine and xenon concentrations, regulating rod (RR) positions, exit quality and downcomer enthalpy.

$\bar{\sigma}_{Xi} = \sigma_{Xi}/E_e \Sigma_{fi} V_i$, $\hat{k}_2 = h_s/h_c$, and $\hat{k}_1 = h_w \hat{k}_2$. $v_k$ is control signal applied to the RR drive. Other notations have their standard meanings (Shimjith et al. 2011). Values of $e_{vx_i}$, $e_{x_i}$ and other constants are specified in Shimjith et al. (2011). The overall reactivity is denoted by $\rho_i$ in (3.1) and is defined as $\rho_i = \rho_{i_u} + \rho_{iX} + \rho_{i_x}$, where $\rho_{i_u}$, $\rho_{iX}$ and $\rho_{i_x}$ are the reactivity feedbacks due to the control rods, xenon and coolant void fraction respectively. Linearizing (3.1)–(3.7) around the steady-state, a standard state-space representation can be acquired by defining a state vector

$$\mathbf{z} = \left[\mathbf{z}_H^T \; \mathbf{z}_X^T \; \mathbf{z}_I^T \; \delta h_d \; \mathbf{z}_C^T \; \mathbf{z}_x^T \; \mathbf{z}_P^T\right]^T, \tag{3.8}$$

where $\mathbf{z}_H = \left[\delta H_2 \; \delta H_4 \; \delta H_6 \; \delta H_8\right]^T$ and the rest $\mathbf{z}_\xi = \left[(\delta\xi_1/\xi_{1_0}) \cdots (\delta\xi_{17}/\xi_{17_0})\right]^T$, $\xi = X, I, C, x, P$, in which $\delta$ stands for the deviation from respective steady-state value. Similarly, input and output vectors are expounded as

$$\mathbf{u} = \left[\delta v_2 \; \delta v_4 \; \delta v_6 \; \delta v_8\right]^T, \tag{3.9}$$

$$\mathbf{y} = \left[y_t \; y_1 \cdots y_{17}\right]^T, \tag{3.10}$$

where $y_t = \sum_{i=1}^{17} \frac{\delta P_i}{\sum_{j=1}^{17} P_{j0}}$ and $y_i = \frac{\delta P_i}{P_{i_0}}$ are respectively normalized total reactor power and nodal powers, in which $P_{i_0}$ is a steady-state value of nodal power. Eventually, the system (3.1)–(3.7) is expressed as

$$\dot{\mathbf{z}} = \hat{\mathbf{A}}z + \mathbf{B}\mathbf{u} + \mathbf{B}_f \delta q_f, \tag{3.11}$$

$$\mathbf{y} = \mathbf{C}\mathbf{z}, \tag{3.12}$$

where $q_f$ is the flow rate of feed water. It is established that this system is observable and controllable. Further, it is seen that the system matrix $\hat{\mathbf{A}}$ eigenvalues fall into three clusters that are broadly separated from one another. There are 38 'slow' eigenvalues in the first cluster ranging from $-1.8870 \times 10^{-4}$ to $7.455 \times 10^{-3}$. The third cluster has 17 'very fast' eigenvalues from $-7.2516$ to $-2.7626 \times 10^{2}$, while the second cluster has 35 'fast' eigenvalues, from $-1.2514 \times 10^{-2}$ to $-1.8395 \times 10^{-1}$. The fast and very fast clusters' eigenvalues are all confirmed to be in the left half of the $s$-plane, while the slow cluster has six in the right half and four at the origin. This suggests instability, necessitating the use of a capable spatial power feedback controller for the AHWR's stable and reliable operation. Large nuclear reactors are often controlled by controlling both total and spatial power, i.e.,

$$\mathbf{u} = \mathbf{u}_t + \mathbf{u}_s, \tag{3.13}$$

where $\mathbf{u}_t$ and $\mathbf{u}_s$ are respectively total and spatial power feedback components. Thus, before designing any spatial power controller, total power control is established. The total power control of AHWR is achieved by generating control signals for RRs, based on total power feedback as

$$\mathbf{u}_t = -\mathbf{K}\mathbf{y}, \tag{3.14}$$

where (4 × 18) order matrix $\mathbf{K} = \left[\mathbf{k}_t\ \mathbf{0} \cdots \mathbf{0}\right]$. Here, $\mathbf{0}$ is a column vector of order 4 and $\mathbf{k}_t = \left[k_t\ k_t\ k_t\ k_t\right]^T$, such that, the feedback gain corresponding to total power is $k_t$ for all RRs and is zero corresponding to nodal powers. Using (3.12)–(3.14), the state Eq. (3.11) changes to

$$\dot{\mathbf{z}} = \mathbf{A}\mathbf{z} + \mathbf{B}\mathbf{u}_s + \mathbf{B}_f \delta q_f, \tag{3.15}$$

where $\mathbf{A} = (\hat{\mathbf{A}} - BKC)$. Setting k = 12.5, the total power of the reactor is successfully controlled. Even after the total power feedback, few things remain unchanged, such as the controllability and observability of the system (**A**, **B**, **C**), three groups of eigenvalues of the same orders as those of $\hat{\mathbf{A}}$, and position of eigenvalues in the fast and very fast groups. More importantly, three eigenvalues at the origin and four eigenvalues with positive real parts can still be seen in a slow cluster. Because of that, even if the total power in the reactor is controlled with input (3.14), the power in the different areas of the reactor core may undergo variations and that being so, spatial feedback control is indispensable.

In the case of AHWR, out-of-core ion chambers and in-core detectors are used to calculate the neutron flux. Ion chambers operating in the low power range and in-core detectors are used to estimate the reactor's total power (Sinha and Kakodkar 2006). Two key issues must be addressed to apply a state feedback-based spatial power controller for AHWR, namely obtaining estimates of immeasurable states by an observer, and constructing an observer for the higher order system. In other words, an observer is required to precisely determine some state variables because sensors are unable to monitor some of them, such as the concentrations of xenon, iodine, and precursors. On the other hand, it would be computationally difficult to construct an observer for a high-dimensional system with state variables moving at widely different speeds. The observers proposed in this chapter overcome the stated difficulty.

## 3.3 Observers for Three-Time-Scale System

Consider a linear time-invariant three-time-scale system consisting of $\mathbf{z}_1 \in \Re^{n_1}$ slow, $\mathbf{z}_2 \in \Re^{n_2}$ fast and $\mathbf{z}_3 \in \Re^{n_3}$ very fast states as

$$\left.\begin{aligned}
\dot{\mathbf{z}}_1 &= \mathbf{A}_{11}\mathbf{z}_1 + \mathbf{A}_{12}\mathbf{z}_2 + \mathbf{A}_{13}\mathbf{z}_3 + \mathbf{B}_1\mathbf{u},\\
\varepsilon\dot{\mathbf{z}}_2 &= \mathbf{A}_{21}\mathbf{z}_1 + \mathbf{A}_{22}\mathbf{z}_2 + \mathbf{A}_{23}\mathbf{z}_3 + \mathbf{B}_2\mathbf{u},\\
\mu\dot{\mathbf{z}}_3 &= \mathbf{A}_{31}\mathbf{z}_1 + \mathbf{A}_{32}\mathbf{z}_2 + \mathbf{A}_{33}\mathbf{z}_3 + \mathbf{B}_3\mathbf{u},\\
\mathbf{y} &= \mathbf{C}_1\mathbf{z}_1 + \mathbf{C}_2\mathbf{z}_2 + \mathbf{C}_3\mathbf{z}_3,
\end{aligned}\right\} \tag{3.16}$$

with $n_1 + n_2 + n_3 = n$. $\mathbf{u} \in \Re^m$, $\mathbf{y} \in \Re^p$ and let $\mathbf{z} = \left[\mathbf{z}_1^T\ \mathbf{z}_2^T\ \mathbf{z}_3^T\right]^T$,

$$\mathbf{A} = \begin{bmatrix} \mathbf{A}_{11} & \mathbf{A}_{12} & \mathbf{A}_{13} \\ \frac{\mathbf{A}_{21}}{\varepsilon} & \frac{\mathbf{A}_{22}}{\varepsilon} & \frac{\mathbf{A}_{23}}{\varepsilon} \\ \frac{\mathbf{A}_{31}}{\mu} & \frac{\mathbf{A}_{32}}{\mu} & \frac{\mathbf{A}_{33}}{\mu} \end{bmatrix}, \mathbf{B} = \begin{bmatrix} \mathbf{B}_1 \\ \frac{\mathbf{B}_2}{\varepsilon} \\ \frac{\mathbf{B}_3}{\mu} \end{bmatrix}, \mathbf{C}^T = \begin{bmatrix} \mathbf{C}_1^T \\ \mathbf{C}_2^T \\ \mathbf{C}_3^T \end{bmatrix}.$$

Matrices $\mathbf{A}_{ij}$, $\mathbf{B}_i$ and $\mathbf{C}_i$ are of proper matrix dimensions. $\varepsilon$ and $\mu$ state speed ratios of $\mathbf{z}_1$ versus $\mathbf{z}_2$ and $\mathbf{z}_1$ versus $\mathbf{z}_3$ states respectively, such that $\mu \ll \varepsilon \ll 1$. More precisely, the system (3.16) has three separate groups of eigenvalues in which $n_1$ eigenvalues are closed to the origin, and $n_2$ and $n_3$ eigenvalues are far and farther from the origin. If the eigenvalues of $\mathbf{A}$ are arranged in the increasing order of absolute values as

$$\psi(\mathbf{A}) = (\psi_1, \ldots, \psi_{n_1}, \psi_{n_1+1}, \ldots, \psi_{n_1+n_2}, \psi_{n_1+n_2+1}, \ldots, \psi_n),$$

where $\psi_i$ is $i$th eigenvalue with

$$0 \leq |\psi_1| < \cdots < |\psi_{n_1}| \ll |\psi_{n_1+1}| < \cdots$$
$$< |\psi_{n_1+n_2}| \ll |\psi_{n_1+n_2+1}| < \cdots < |\psi_n|,$$

then referring (Naidu 1988)

$$\varepsilon = \frac{|\psi_{n_1}|}{|\psi_{n_1+1}|} \text{ and } \mu = \frac{|\psi_{n_1}|}{|\psi_{n_1+n_2+1}|}.$$

Before moving ahead, the following assumptions are imposed on (3.16).

**Assumption 3.1** The pair $(\mathbf{A}, \mathbf{B})$ is controllable and pair $(\mathbf{A}, \mathbf{C})$ is observable.

**Assumption 3.2** Output matrix $\mathbf{C}$ of every reformulated system, considered for reduced-order observer design has a full rank, i.e., rank equal to outputs accessible for measurement.

**Assumption 3.3** Matrices $\mathbf{A}_{22}$ and $\mathbf{A}_{33}$ are invertible (Naidu 1988).

To ascertain observer gains and satisfy the observability requirement of the reduced-order subsystems, Assumptions 3.1 and 3.2 are required. To find the solutions to algebraic equations, it is necessary to make the crucial Assumption 3.3 in time-scale theory. The recommended full- and reduced-order observers for the system (3.16) are given in the ensuing subsections.

### *3.3.1 Full-Order Observer Design*

The mathematical difficulty in computing observer gain for a numerically ill-conditioned system (3.16) is effectively reduced by a three-stage design. This is explained using the duality principle. As the transpose of the observer system, i.e., $(\mathbf{A}^T - \mathbf{C}^T\mathbf{L}^T)$ produces the dual form to the original system $(\mathbf{A} - \mathbf{BK})$, the similar method can be used to design feedback matrices for $\dot{\mathbf{z}} = (\mathbf{A} - \mathbf{BK})\mathbf{z}$ and $\dot{\boldsymbol{\Theta}} = (\mathbf{A} - \mathbf{LC})\hat{\mathbf{z}} + \mathbf{Ly}$. Let us consider an analogous system to (3.16) as

$$\begin{bmatrix} \dot{\mathbf{x}}_1 \\ \dot{\mathbf{x}}_2 \\ \dot{\mathbf{x}}_3 \end{bmatrix} = \begin{bmatrix} \mathbf{A}_{11}^T & \frac{\mathbf{A}_{21}^T}{\varepsilon} & \frac{\mathbf{A}_{31}^T}{\mu} \\ \mathbf{A}_{12}^T & \frac{\mathbf{A}_{22}^T}{\varepsilon} & \frac{\mathbf{A}_{32}^T}{\mu} \\ \mathbf{A}_{13}^T & \frac{\mathbf{A}_{23}^T}{\varepsilon} & \frac{\mathbf{A}_{33}^T}{\mu} \end{bmatrix} \begin{bmatrix} \mathbf{x}_1 \\ \mathbf{x}_2 \\ \mathbf{x}_3 \end{bmatrix} + \begin{bmatrix} \mathbf{C}_1^T \\ \mathbf{C}_2^T \\ \mathbf{C}_3^T \end{bmatrix} \nu. \tag{3.17}$$

Introducing

$$\begin{bmatrix} \mathbf{q}_1 \\ \mathbf{q}_2 \\ \mathbf{q}_3 \end{bmatrix} = \begin{bmatrix} \mathbf{I}_{n1} & \mathbf{0} & \mathbf{0} \\ \mathbf{0} & \frac{\mathbf{I}_{n2}}{\varepsilon} & \mathbf{0} \\ \mathbf{0} & \mathbf{0} & \frac{\mathbf{I}_{n3}}{\mu} \end{bmatrix} \begin{bmatrix} \mathbf{x}_1 \\ \mathbf{x}_2 \\ \mathbf{x}_3 \end{bmatrix} = \mathbf{T}^T \begin{bmatrix} \mathbf{x}_1 \\ \mathbf{x}_2 \\ \mathbf{x}_3 \end{bmatrix}, \tag{3.18}$$

system (3.17) is transformed into an explicitly singularly perturbed form

$$\begin{bmatrix} \dot{\mathbf{q}}_1 \\ \dot{\mathbf{q}}_2 \\ \dot{\mathbf{q}}_3 \end{bmatrix} = \begin{bmatrix} \mathbf{A}_{11}^T & \mathbf{A}_{21}^T & \mathbf{A}_{31}^T \\ \frac{\mathbf{A}_{12}^T}{\varepsilon} & \frac{\mathbf{A}_{22}^T}{\varepsilon} & \frac{\mathbf{A}_{32}^T}{\varepsilon} \\ \frac{\mathbf{A}_{13}^T}{\mu} & \frac{\mathbf{A}_{23}^T}{\mu} & \frac{\mathbf{A}_{33}^T}{\mu} \end{bmatrix} \begin{bmatrix} \mathbf{q}_1 \\ \mathbf{q}_2 \\ \mathbf{q}_3 \end{bmatrix} + \begin{bmatrix} \mathbf{C}_1^T \\ \frac{\mathbf{C}_2^T}{\varepsilon} \\ \frac{\mathbf{C}_3^T}{\mu} \end{bmatrix} \nu, \tag{3.19}$$

where $\mathbf{I}_n$ is an identity matrix of order $n$, $\mathbf{q}_1 \in \Re^{n_1}$, $\mathbf{q}_2 \in \Re^{n_2}$ and $\mathbf{q}_3 \in \Re^{n_3}$. For designing an observer in three stages, first of all, the system (3.19) is changed to a lower triangular form and then observer gains are designed independently for slow, fast, and very fast subsystems by two-time-scale similarity transformations. Now, regrouping states of (3.19) as $\mathbf{q}_a = \left[\mathbf{q}_1^T\ \mathbf{q}_2^T\right]^T$ and $\mathbf{q}_b = \mathbf{q}_3$ result in

$$\begin{bmatrix} \dot{\mathbf{q}}_a \\ \dot{\mathbf{q}}_b \end{bmatrix} = \begin{bmatrix} \bar{\mathbf{A}}_{11}^T & \bar{\mathbf{A}}_{21}^T \\ \frac{\bar{\mathbf{A}}_{12}^T}{\mu} & \frac{\bar{\mathbf{A}}_{22}^T}{\mu} \end{bmatrix} \begin{bmatrix} \mathbf{q}_a \\ \mathbf{q}_b \end{bmatrix} + \begin{bmatrix} \bar{\mathbf{C}}_1^T \\ \frac{\bar{\mathbf{C}}_2^T}{\mu} \end{bmatrix} \nu, \tag{3.20}$$

where $\bar{\mathbf{A}}_{ij}$ and $\bar{\mathbf{C}}_i$ are updated accordingly. The system (3.20) is a two-time-scale representation of (3.19), where $n_1$ 'slow' and $n_2$ 'fast' states are merged to form $(n_1 + n_2)$ 'slow' states and $n_3$ 'very fast' states are taken as 'fast' states. Now, applying

$$\mathbf{q}_s = \mathbf{q}_a + \mu\mathbf{P}^T\mathbf{q}_b, \tag{3.21}$$

to the system (3.20), produces

$$\begin{bmatrix} \dot{\mathbf{q}}_s \\ \dot{\mathbf{q}}_b \end{bmatrix} = \begin{bmatrix} \bar{\mathbf{A}}_s^T & \mathbf{0} \\ \frac{\bar{\mathbf{A}}_{12}^T}{\mu} & \frac{\bar{\mathbf{A}}_f^T}{\mu} \end{bmatrix} \begin{bmatrix} \mathbf{q}_s \\ \mathbf{q}_b \end{bmatrix} + \begin{bmatrix} \bar{\mathbf{C}}_s^T \\ \frac{\bar{\mathbf{C}}_2^T}{\mu} \end{bmatrix} \nu, \tag{3.22}$$

where $\bar{\mathbf{A}}_s^T = \bar{\mathbf{A}}_{11}^T + \mathbf{P}^T \bar{\mathbf{A}}_{12}^T$, $\bar{\mathbf{A}}_f^T = \bar{\mathbf{A}}_{22}^T - \mu \bar{\mathbf{A}}_{12}^T \mathbf{P}^T$, $\bar{\mathbf{C}}_s = \bar{\mathbf{C}}_1^T + \mathbf{P}^T \bar{\mathbf{C}}_2^T$ and

$$\bar{\mathbf{A}}_{21}^T + \mathbf{P}^T \bar{\mathbf{A}}_{22}^T - \mu(\bar{\mathbf{A}}_{11}^T + \mathbf{P}^T \bar{\mathbf{A}}_{12}^T)\mathbf{P}^T = 0. \tag{3.23}$$

The unique solution of $\mathbf{P}^T$ is gained by solving (3.23) by fixed point iteration method, using first-order approximation as $\mathbf{P}^{T^{(0)}} = -\bar{\mathbf{A}}_{21}^T \bar{\mathbf{A}}_{22}^{-T}$, for sufficiently small value of $\mu$ under Assumption 3.3 (Naidu 1988). Accordingly, (3.22) is modified to lower triangular form, where $\mathbf{q}_s$ is completely decoupled from $\mathbf{q}_b$. Recall that, $\mathbf{q}_s$ combines $n_1$ 'slow' and $n_2$ 'fast' states of the system (3.19). Due to this, (3.22) can also be written in two-time-scale form, by partitioning matrices $\bar{\mathbf{A}}_s^T$ and $\bar{\mathbf{C}}_s^T$ suitably, as

$$\begin{bmatrix} \dot{\mathbf{q}}_{s1} \\ \dot{\mathbf{q}}_{s2} \end{bmatrix} = \begin{bmatrix} \bar{\mathbf{A}}_{s11}^T & \bar{\mathbf{A}}_{s21}^T \\ \frac{\bar{\mathbf{A}}_{s12}^T}{\varepsilon} & \frac{\bar{\mathbf{A}}_{s22}^T}{\varepsilon} \end{bmatrix} \begin{bmatrix} \mathbf{q}_{s1} \\ \mathbf{q}_{s2} \end{bmatrix} + \begin{bmatrix} \bar{\mathbf{C}}_{s1}^T \\ \frac{\bar{\mathbf{C}}_{s2}^T}{\varepsilon} \end{bmatrix} \nu, \tag{3.24}$$

where $\mathbf{q}_s = \left[\mathbf{q}_{s1}^T \ \mathbf{q}_{s2}^T\right]^T$ with $\mathbf{q}_{s1} \in \Re^{n_1}$ and $\mathbf{q}_{s2} \in \Re^{n_2}$. At this moment, a linear transformation

$$\mathbf{q}_{ss} = \mathbf{q}_{s1} + \varepsilon \mathbf{P}_s^T \mathbf{q}_{s2}, \tag{3.25}$$

is applied to (3.24). Here, matrix $\mathbf{P}_s^T$ is reckoned by solving

$$\bar{\mathbf{A}}_{s21}^T + \mathbf{P}_s^T \bar{\mathbf{A}}_{s22}^T - \varepsilon(\bar{\mathbf{A}}_{s11}^T + \mathbf{P}_s^T \bar{\mathbf{A}}_{s12}^T)\mathbf{P}_s^T = 0. \tag{3.26}$$

Solution of $\mathbf{P}_s^T$ can be obtained under Assumption 3.3 for a small value of $\varepsilon$ in the same way as that of $\mathbf{P}^T$ in (3.23). Resultantly, a lower triangular form of (3.24) is reached

$$\begin{bmatrix} \dot{\mathbf{q}}_{ss} \\ \dot{\mathbf{q}}_{s2} \end{bmatrix} = \begin{bmatrix} \bar{\mathbf{A}}_{ss}^T & \mathbf{0} \\ \frac{\bar{\mathbf{A}}_{s12}^T}{\varepsilon} & \frac{\bar{\mathbf{A}}_{sf}^T}{\varepsilon} \end{bmatrix} \begin{bmatrix} \mathbf{q}_{s1} \\ \mathbf{q}_{s2} \end{bmatrix} + \begin{bmatrix} \bar{\mathbf{C}}_{ss}^T \\ \frac{\bar{\mathbf{C}}_{s2}^T}{\varepsilon} \end{bmatrix} \nu, \tag{3.27}$$

where $\bar{\mathbf{A}}_{ss}^T = \bar{\mathbf{A}}_{s11}^T + \mathbf{P}_s^T \bar{\mathbf{A}}_{s12}^T$, $\bar{\mathbf{A}}_{sf}^T = \bar{\mathbf{A}}_{s22}^T - \varepsilon \bar{\mathbf{A}}_{s12}^T \mathbf{P}_s^T$ and $\bar{\mathbf{C}}_{ss}^T = \bar{\mathbf{C}}_{s1}^T + \mathbf{P_s}^T \bar{\mathbf{C}}_{s2}^T$. In such a way, using block-triangular forms (3.22) and (3.27) of two-time-scale systems, the original system (3.19) is converted into block-triangular form

$$\begin{bmatrix} \dot{\mathbf{q}}_{ss} \\ \dot{\mathbf{q}}_{s2} \\ \dot{\mathbf{q}}_3 \end{bmatrix} = \begin{bmatrix} \bar{\mathbf{A}}_{ss}^T & \mathbf{0} & \mathbf{0} \\ \frac{\bar{\mathbf{A}}_{s12}^T}{\varepsilon} & \frac{\bar{\mathbf{A}}_{sf}^T}{\varepsilon} & \mathbf{0} \\ \frac{\bar{\mathbf{A}}_{13}^T}{\mu} & \frac{\bar{\mathbf{A}}_{23}^T}{\mu} & \frac{\bar{\mathbf{A}}_f^T}{\mu} \end{bmatrix} \begin{bmatrix} \mathbf{q}_{ss} \\ \mathbf{q}_{s2} \\ \mathbf{q}_3 \end{bmatrix} + \begin{bmatrix} \bar{\mathbf{C}}_{ss}^T \\ \frac{\bar{\mathbf{C}}_{s2}^T}{\varepsilon} \\ \frac{\bar{\mathbf{C}}_2^T}{\mu} \end{bmatrix} \nu, \tag{3.28}$$

where $\bar{\mathbf{A}}_{12}^T$ is appropriately partitioned as $[\bar{\mathbf{A}}_{13}^T \quad \bar{\mathbf{A}}_{23}^T]$. The form (3.28) also satisfies the eigenvalue clustering property of three-time-scale system discussed earlier, i.e., subsystems $\bar{\mathbf{A}}_{ss}^T$, $\frac{\bar{\mathbf{A}}_{sf}^T}{\varepsilon}$ and $\frac{\bar{\mathbf{A}}_{f}^T}{\mu}$ include respectively $n_1$ slow, $n_2$ fast and $n_3$ very fast eigenvalues, such that

$$max|\psi(\bar{\mathbf{A}}_{ss}^T)| \ll min\left|\psi\left(\frac{\bar{\mathbf{A}}_{sf}^T}{\varepsilon}\right)\right|,$$

$$max\left|\psi\left(\frac{\bar{\mathbf{A}}_{sf}^T}{\varepsilon}\right)\right| \ll min\left|\psi\left(\frac{\bar{\mathbf{A}}_{f}^T}{\mu}\right)\right|.$$

Now, one can proceed for a three-stage observer gain design starting with $\mathbf{q}_{ss}$, i.e., the slow subsystem. To begin with, the input is designed for the system (3.27) and then for (3.22). In the first stage, applying

$$\nu = -\mathbf{L}_{ss}^T\mathbf{q}_{ss} + \nu_{sf}, \tag{3.29}$$

to the system (3.27), leads to

$$\begin{bmatrix} \dot{\mathbf{q}}_{ss} \\ \dot{\mathbf{q}}_{s2} \end{bmatrix} = \begin{bmatrix} \bar{\mathbf{A}}_{ss}^T - \bar{\mathbf{C}}_{ss}^T\mathbf{L}_{ss}^T & \mathbf{0} \\ \frac{\bar{\mathbf{A}}_{s12}^T - \bar{\mathbf{C}}_{s2}^T\mathbf{L}_{ss}^T}{\varepsilon} & \frac{\bar{\mathbf{A}}_{sf}^T}{\varepsilon} \end{bmatrix} \begin{bmatrix} \mathbf{q}_{ss} \\ \mathbf{q}_{s2} \end{bmatrix} + \begin{bmatrix} \bar{\mathbf{C}}_{ss}^T \\ \frac{\bar{\mathbf{C}}_{s2}^T}{\varepsilon} \end{bmatrix} \nu_{sf}, \tag{3.30}$$

where $\mathbf{L}_{ss}^T$ is selected, such that $\psi(\bar{\mathbf{A}}_{ss}^T - \bar{\mathbf{C}}_{ss}^T\mathbf{L}_{ss}^T) = \psi_s^{desired}$, where subscript '$s$' indicates slow subsystem. As the similarity transformations map the original system from one form to another without the loss of controllability (or observability), the pair $(\bar{\mathbf{A}}_{ss}^T, \bar{\mathbf{C}}_{ss}^T)$ is controllable (or the pair $(\bar{\mathbf{A}}_{ss}, \bar{\mathbf{C}}_{ss})$ is observable), due to which an arbitrary pole placement is possible under Assumption 3.1. Now, using

$$\mathbf{q}_{sf} = -\mathbf{M}_s^T\mathbf{q}_{ss} + \mathbf{q}_{s2}, \tag{3.31}$$

system (3.30) is modified to

$$\begin{bmatrix} \dot{\mathbf{q}}_{ss} \\ \dot{\mathbf{q}}_{sf} \end{bmatrix} = \begin{bmatrix} \bar{\mathbf{A}}_{ss}^T - \bar{\mathbf{C}}_{ss}^T\mathbf{L}_{ss}^T & \mathbf{0} \\ \mathbf{0} & \frac{\bar{\mathbf{A}}_{sf}^T}{\varepsilon} \end{bmatrix} \begin{bmatrix} \mathbf{q}_{ss} \\ \mathbf{q}_{sf} \end{bmatrix} + \begin{bmatrix} \bar{\mathbf{C}}_{ss}^T \\ \frac{\bar{\mathbf{C}}_{sf}^T}{\varepsilon} \end{bmatrix} \nu_{sf}, \tag{3.32}$$

where $\bar{\mathbf{C}}_{sf}^T = \bar{\mathbf{C}}_{s2}^T - \varepsilon\mathbf{M}_s^T\bar{\mathbf{C}}_{ss}^T$ and

$$\bar{\mathbf{A}}_{s12}^T - \bar{\mathbf{C}}_{s2}^T\mathbf{L}_{ss}^T - \varepsilon\mathbf{M}_s^T(\bar{\mathbf{A}}_{ss}^T - \bar{\mathbf{C}}_{ss}^T\mathbf{L}_{ss}^T) + \bar{\mathbf{A}}_{sf}^T\mathbf{M}_s^T = 0. \tag{3.33}$$

Matrix $\mathbf{M}_s^T$ can be determined from (3.33) using fixed point iteration (with $\mathbf{M}_s^{T^{(0)}} = -\bar{\mathbf{A}}_{sf}^{-T}(\bar{\mathbf{A}}_{s12}^T - \bar{\mathbf{C}}_{s2}^T\mathbf{L}_{ss}^T)$ as the first-order approximation) and following assumption

**Assumption 3.4** $\psi(\bar{\mathbf{A}}_{ss}^T - \bar{\mathbf{C}}_{ss}^T\mathbf{L}_{ss}^T) \neq \psi\left(\frac{\bar{\mathbf{A}}_{sf}^T}{\varepsilon}\right)$.

Existence of $\bar{\mathbf{A}}_{sf}^{-T}$ is ensured by Assumption 3.3 and Assumption 3.4 is satisfied because $(\bar{\mathbf{A}}_{ss}^T - \bar{\mathbf{C}}_{ss}^T\mathbf{L}_{ss}^T)$ is designed to have desired 'slow' eigenvalues, while $\frac{\bar{\mathbf{A}}_{sf}^T}{\varepsilon}$ has original subsystem 'fast' eigenvalues. In (3.32), subsystem $\mathbf{q}_{sf}$ is made totally independent of $\mathbf{q}_{ss}$. Thereupon, in the second stage

$$\nu_{sf} = -\mathbf{L}_{sf}^T\mathbf{q}_{sf}, \tag{3.34}$$

is passed to the system (3.32) as

$$\begin{bmatrix} \dot{\mathbf{q}}_{ss} \\ \dot{\mathbf{q}}_{sf} \end{bmatrix} = \begin{bmatrix} \bar{\mathbf{A}}_{ss}^T - \bar{\mathbf{C}}_{ss}^T\mathbf{L}_{ss}^T & -\bar{\mathbf{C}}_{ss}^T\mathbf{L}_{sf}^T \\ \mathbf{0} & \frac{\bar{\mathbf{A}}_{sf}^T - \bar{\mathbf{C}}_{sf}^T\mathbf{L}_{sf}^T}{\varepsilon} \end{bmatrix} \begin{bmatrix} \mathbf{q}_{ss} \\ \mathbf{q}_{sf} \end{bmatrix}, \tag{3.35}$$

so that $\psi\left(\frac{\bar{\mathbf{A}}_{sf}^T - \bar{\mathbf{C}}_{sf}^T\mathbf{L}_{sf}^T}{\varepsilon}\right) = \psi_f^{desired}$. The subscript '$f$' stands for fast subsystem. The controllability of $\left(\frac{\bar{\mathbf{A}}_{sf}^T}{\varepsilon}, \frac{\bar{\mathbf{C}}_{sf}^T}{\varepsilon}\right)$ is preserved due to Assumption 3.1. Thus, input for the system (3.24) is computed using (3.25), (3.29), (3.31) and (3.34) as

$$\nu = -\mathbf{L}_s^T\left[\mathbf{q}_{s1}^T \; \mathbf{q}_{s2}^T\right]^T = -\mathbf{L}_s^T\mathbf{q}_s, \tag{3.36}$$

where $\mathbf{L}_s^T = \left[\mathbf{L}_{ss}^T - \mathbf{L}_{sf}^T\mathbf{M}_s^T \;\; \varepsilon(\mathbf{L}_{ss}^T - \mathbf{L}_{sf}^T\mathbf{M}_s^T)\mathbf{P}_s^T + \mathbf{L}_{sf}^T\right]$. Actually, (3.36) is the input for $\mathbf{q}_s$, which is the decoupled part of (3.22). Therefore, the overall input $\nu$ has the form

$$\nu = -\mathbf{L}_s^T\mathbf{q}_s + \nu_f. \tag{3.37}$$

Passing on (3.37) to (3.22) and changing coordinates as

$$\mathbf{q}_f = -\mathbf{M}^T\mathbf{q}_s + \mathbf{q}_b, \tag{3.38}$$

the subsystem $\mathbf{q}_f$ is isolated from $\mathbf{q}_s$ as

$$\begin{bmatrix} \dot{\mathbf{q}}_s \\ \dot{\mathbf{q}}_f \end{bmatrix} = \begin{bmatrix} \bar{\mathbf{A}}_s^T - \bar{\mathbf{C}}_s^T\mathbf{L}_s^T & \mathbf{0} \\ \mathbf{0} & \frac{\bar{\mathbf{A}}_f^T}{\mu} \end{bmatrix} \begin{bmatrix} \mathbf{q}_s \\ \mathbf{q}_f \end{bmatrix} + \begin{bmatrix} \bar{\mathbf{C}}_s^T \\ \frac{\bar{\mathbf{C}}_f^T}{\mu} \end{bmatrix} \nu_f, \tag{3.39}$$

where $\bar{\mathbf{C}}_f^T = \bar{\mathbf{C}}_2^T - \mu\mathbf{M}^T\bar{\mathbf{C}}_s^T$ and $\mathbf{M}^T$ is evaluated by setting

$$\bar{\mathbf{A}}_{12}^T - \bar{\mathbf{C}}_2^T\mathbf{L}_s^T - \mu\mathbf{M}^T(\bar{\mathbf{A}}_s^T - \bar{\mathbf{C}}_s^T\mathbf{L}_s^T) + \bar{\mathbf{A}}_f^T\mathbf{M}^T = 0, \tag{3.40}$$

similar to (3.33) with one more assumption:

**Assumption 3.5** $\psi(\bar{\mathbf{A}}_s^T - \bar{\mathbf{C}}_s^T\mathbf{L}_s^T) \neq \psi\left(\frac{\bar{\mathbf{A}}_f^T}{\mu}\right)$.

It can be convinced by designing $(\bar{\mathbf{A}}_s^T - \bar{\mathbf{C}}_s^T\mathbf{L}_s^T)$ with desired eigenvalues laying in 'slow' and 'fast' modes. And, $\frac{\bar{\mathbf{A}}_f^T}{\mu}$ has 'very fast' eigenvalues of (3.16). Finally, in the third stage

$$\nu_f = -\mathbf{L}_f^T\mathbf{q}_f, \tag{3.41}$$

is applied to the system (3.39) as

$$\begin{bmatrix} \dot{\mathbf{q}}_s \\ \dot{\mathbf{q}}_f \end{bmatrix} = \begin{bmatrix} \bar{\mathbf{A}}_s^T - \bar{\mathbf{C}}_s^T\mathbf{L}_s^T & -\bar{\mathbf{C}}_s^T\mathbf{L}_f^T \\ \mathbf{0} & \frac{\bar{\mathbf{A}}_f^T - \bar{\mathbf{C}}_f^T\mathbf{L}_f^T}{\mu} \end{bmatrix} \begin{bmatrix} \mathbf{q}_s \\ \mathbf{q}_f \end{bmatrix}, \tag{3.42}$$

so as to assign $\psi\left(\frac{\bar{\mathbf{A}}_f^T - \bar{\mathbf{C}}_f^T\mathbf{L}_f^T}{\mu}\right) = \psi_{ff}^{desired}$, assuming the controllability of $\left(\frac{\bar{\mathbf{A}}_f^T}{\mu}, \frac{\bar{\mathbf{C}}_f^T}{\mu}\right)$ fulfilled by Assumption 3.1, where subscript '$ff$' denotes very fast subsystem. The complete input for the system (3.17), obtained from (3.37) and (3.41), has the form

$$\nu = -\mathbf{L}^T\mathbf{x}, \tag{3.43}$$

where $\mathbf{x} = \left[\mathbf{x}_1^T\ \mathbf{x}_2^T\ \mathbf{x}_3^T\right]^T$ and $\mathbf{L}^T = \bar{\mathbf{L}}^T\mathbf{T}^{-T}$ in which

$$\bar{\mathbf{L}}^T = \left[\mathbf{L}_s^T - \mathbf{L}_f^T\mathbf{M}^T\ \ \mu(\mathbf{L}_s^T - \mathbf{L}_f^T\mathbf{M}^T)\mathbf{P}^T + \mathbf{L}_f^T\right],$$

and $\mathbf{T}^T$ is given by (3.18). This is the composite feedback gain, $\mathbf{L}$ derived from subsystem feedback gains designed independently in three different stages using well-conditioned matrices, so that the original system (3.17) is asymptotically stable, i.e., $\psi(\mathbf{A^T} - \mathbf{C^T}\mathbf{L^T}) < 0$. Uniqueness in this three-stage design and subsequent advantages are summarized below.

1. The lower triangular form (3.28) of (3.19) is attained by solving only two non-symmetrical Riccati equations (3.23) and (3.26) for $\mathbf{P}$ and $\mathbf{P}_s$ respectively and three-stage design is performed by solving two Sylvester equations (3.33) and (3.40) respectively for $\mathbf{M}_s$ and $\mathbf{M}$, i.e., only four equations are required to be solved compared to six in Radisavljevic-Gajic et al. (2017), which reduced online computation time.
2. The computations are performed using a two-time-scale representation rather than a three-time-scale one, which significantly reduces computational complexity and offline computation time without sacrificing accuracy and gives the option to design unique observer gains for local subsystems, such as slow, fast, and very fast subsystems.

### 3.3.2 Reduced-Order Observer Design

#### 3.3.2.1 Estimation of Slow and Fast States

In this case, it is assumed that $\mathbf{z}_1$ and $\mathbf{z}_2$ are to be estimated by an observer and $\mathbf{z}_3$ is to be determined from the outputs. Because of this, output $\mathbf{y} \in \Re^{n_3}$ implying $rank(\mathbf{C}) = n_3$ as per Assumption 3.2. For the same, system (3.16) is redefined as

$$\left.\begin{aligned} \dot{\mathbf{z}}_{sf_1} &= \mathbf{A}_{sf_{11}}\mathbf{z}_{sf_1} + \mathbf{A}_{sf_{12}}\mathbf{z}_3 + \mathbf{B}_{sf_1}\mathbf{u}, \\ \mu\dot{\mathbf{z}}_3 &= \mathbf{A}_{sf_{21}}\mathbf{z}_{sf_1} + \mathbf{A}_{33}\mathbf{z}_3 + \mathbf{B}_3\mathbf{u}, \\ \mathbf{y} &= \mathbf{z}_3, \end{aligned}\right\} \tag{3.44}$$

where states to be estimated $\mathbf{z}_{sf_1} = \left[\mathbf{z}_1^T \ \mathbf{z}_2^T\right]^T$ and

$$\mathbf{A}_{sf_{11}} = \begin{bmatrix} \mathbf{A}_{11} & \mathbf{A}_{12} \\ \frac{\mathbf{A}_{21}}{\varepsilon} & \frac{\mathbf{A}_{22}}{\varepsilon} \end{bmatrix}, \mathbf{A}_{sf_{12}} = \begin{bmatrix} \mathbf{A}_{13} \\ \frac{\mathbf{A}_{23}}{\varepsilon} \end{bmatrix}, \mathbf{A}_{sf_{21}} = \left[\mathbf{A}_{31} \ \mathbf{A}_{32}\right], \mathbf{B}_{sf_1} = \begin{bmatrix} \mathbf{B}_1 \\ \frac{\mathbf{B}_2}{\varepsilon} \end{bmatrix}.$$

The reduced-order observer (Chen 1999) for (3.44) has the form

$$\dot{\hat{\mathbf{z}}}_{sf} = \mathbf{A}_{sf}\hat{\mathbf{z}}_{sf} + \mathbf{B}_{sf}\mathbf{u} + \mathbf{L}_{sf}\mathbf{y}, \tag{3.45}$$

where $\mathbf{A}_{sf} = \mathbf{A}_{sf_{11}} - \frac{1}{\mu}\mathbf{L}_{sf_1}\mathbf{A}_{sf_{21}}$, $\mathbf{B}_{sf} = \mathbf{B}_{sf_1} - \frac{1}{\mu}\mathbf{L}_{sf_1}\mathbf{B}_3$, $\mathbf{L}_{sf} = \mathbf{A}_{sf}\mathbf{L}_{sf_1} - \frac{1}{\mu}\mathbf{L}_{sf_1}\mathbf{A}_{33} + \mathbf{A}_{sf_{12}}$ and the state estimation is obtained as $\hat{\mathbf{z}}_{sf} = \hat{\mathbf{z}}_{sf_1} - \mathbf{L}_{sf_1}\mathbf{z}_3$. The observer gain matrix $\mathbf{L}_{sf_1}$ is determined, to stabilize (3.45), by two-stage design procedure, analogs to two-stage feedback control design (Radisavljevic-Gajic and Rose 2014). For designing an observer for (3.44), with $\mathbf{z}_{sf_1}$ immeasurable and $\mathbf{z}_3$ measurable states, following dual form is obtained

$$\begin{bmatrix} \dot{\mathbf{x}}_1 \\ \dot{\mathbf{x}}_2 \end{bmatrix} = \begin{bmatrix} \mathbf{A}_{11}^T & \frac{\mathbf{A}_{21}^T}{\varepsilon} \\ \mathbf{A}_{12}^T & \frac{\mathbf{A}_{22}^T}{\varepsilon} \end{bmatrix} \begin{bmatrix} \mathbf{x}_1 \\ \mathbf{x}_2 \end{bmatrix} + \begin{bmatrix} \frac{\mathbf{A}_{31}^T}{\mu} \\ \frac{\mathbf{A}_{32}^T}{\mu} \end{bmatrix} \nu. \tag{3.46}$$

Using $\mathbf{q}_1 = \mu\mathbf{x}_1$ and $\mathbf{q}_2 = \frac{\mu}{\varepsilon}\mathbf{x}_2$, system (3.46) is changed to

$$\begin{bmatrix} \dot{\mathbf{q}}_1 \\ \dot{\mathbf{q}}_2 \end{bmatrix} = \begin{bmatrix} \mathbf{A}_{11}^T & \mathbf{A}_{21}^T \\ \frac{\mathbf{A}_{12}^T}{\varepsilon} & \frac{\mathbf{A}_{22}^T}{\varepsilon} \end{bmatrix} \begin{bmatrix} \mathbf{q}_1 \\ \mathbf{q}_2 \end{bmatrix} + \begin{bmatrix} \mathbf{A}_{31}^T \\ \frac{\mathbf{A}_{32}^T}{\varepsilon} \end{bmatrix} \nu. \tag{3.47}$$

Though (3.46) has $\mu$ in it, the two-time-scale form (3.47) has a speed ratio of $\varepsilon$, which indeed represents that immeasurable states combine slow and fast dynamics and not the very fast one. Applying transformation $\mathbf{q}_s = \mathbf{q}_1 + \varepsilon\mathbf{P}^T\mathbf{q}_2$ to (3.47) and following the procedure from (3.21) to (3.23), generates

$$\begin{bmatrix} \dot{\mathbf{q}}_s \\ \dot{\mathbf{q}}_2 \end{bmatrix} = \begin{bmatrix} \mathbf{A}_s^T & \mathbf{0} \\ \frac{\mathbf{A}_{12}^T}{\varepsilon} & \frac{\mathbf{A}_f^T}{\varepsilon} \end{bmatrix} \begin{bmatrix} \mathbf{q}_s \\ \mathbf{q}_2 \end{bmatrix} + \begin{bmatrix} \mathbf{C}_s^T \\ \frac{\mathbf{A}_{32}^T}{\varepsilon} \end{bmatrix} \nu, \tag{3.48}$$

where $\mathbf{A}_s^T = \mathbf{A}_{11}^T + \mathbf{P}^T \mathbf{A}_{12}^T$, $\mathbf{A}_f^T = \mathbf{A}_{22}^T - \varepsilon \mathbf{A}_{12}^T \mathbf{P}^T$ and $\mathbf{C}_s^T = \mathbf{A}_{31}^T + \mathbf{P}^T \mathbf{A}_{32}$. In the first stage, applying $\nu = -\mathbf{L}_s^T \mathbf{q}_s + \nu_f$ to (3.48) gives

$$\begin{bmatrix} \dot{\mathbf{q}}_s \\ \dot{\mathbf{q}}_2 \end{bmatrix} = \begin{bmatrix} \mathbf{A}_s^T - \mathbf{C}_s^T \mathbf{L}_s^T & \mathbf{0} \\ \frac{\mathbf{A}_{12}^T - \mathbf{A}_{32}^T \mathbf{L}_s^T}{\varepsilon} & \frac{\mathbf{A}_f^T}{\varepsilon} \end{bmatrix} \begin{bmatrix} \mathbf{q}_s \\ \mathbf{q}_2 \end{bmatrix} + \begin{bmatrix} \mathbf{C}_s^T \\ \frac{\mathbf{A}_{32}^T}{\varepsilon} \end{bmatrix} \nu_f, \tag{3.49}$$

where $\mathbf{L}_s^T$ is selected such that $\psi(\mathbf{A}_s^T - \mathbf{C}_s^T \mathbf{L}_s^T) = \psi_s^{desired}$. Later, introducing $\mathbf{q}_f = -\mathbf{M}^T \mathbf{q}_s + \mathbf{q}_2$ for (3.49) and following steps from (3.38) to (3.40) yields

$$\begin{bmatrix} \dot{\mathbf{q}}_s \\ \dot{\mathbf{q}}_f \end{bmatrix} = \begin{bmatrix} \mathbf{A}_s^T - \mathbf{C}_s^T \mathbf{L}_s^T & \mathbf{0} \\ \mathbf{0} & \frac{\mathbf{A}_f^T}{\varepsilon} \end{bmatrix} \begin{bmatrix} \mathbf{q}_s \\ \mathbf{q}_f \end{bmatrix} + \begin{bmatrix} \mathbf{C}_s^T \\ \frac{\mathbf{C}_f^T}{\varepsilon} \end{bmatrix} \nu_f, \tag{3.50}$$

with $\mathbf{C}_f^T = \mathbf{A}_{32}^T - \varepsilon \mathbf{M}^T \mathbf{C}_s^T$. In the second stage, $\nu_f = -\mathbf{L}_f^T \mathbf{q}_f$ is given to the system (3.50) as

$$\begin{bmatrix} \dot{\mathbf{q}}_s \\ \dot{\mathbf{q}}_f \end{bmatrix} = \begin{bmatrix} \mathbf{A}_s^T - \mathbf{C}_s^T \mathbf{L}_s^T & -\mathbf{C}_s^T \mathbf{L}_f^T \\ \mathbf{0} & \frac{\mathbf{A}_f^T - \mathbf{C}_f^T \mathbf{L}_f^T}{\varepsilon} \end{bmatrix} \begin{bmatrix} \mathbf{q}_s \\ \mathbf{q}_f \end{bmatrix}, \tag{3.51}$$

so as to assign $\psi\left(\frac{\mathbf{A}_f^T - \mathbf{C}_f^T \mathbf{L}_f^T}{\varepsilon}\right) = \psi_f^{desired}$. Therefore, the overall input for (3.46) can be rewritten in original states as

$$\nu = -\mathbf{L}_{sf_1}^T \left[\mathbf{x}_1^T \ \mathbf{x}_2^T\right]^T, \tag{3.52}$$

where $\mathbf{L}_{sf_1}^T = \left[\mu(\mathbf{L}_s^T - \mathbf{L}_f^T \mathbf{M}^T) \ \mu(\mathbf{L}_s^T - \mathbf{L}_f^T \mathbf{M}^T)\mathbf{P}^T + \frac{\mu}{\varepsilon}\mathbf{L}_f^T\right]$ is the composite gain, calculated from subsystem gains designed separately, so that the original system (3.46) is stabilized. Estimation error dynamics for slow and fast states are

$$\dot{\mathbf{e}}_{sf} = \left(\mathbf{A}_{sf_{11}} - \frac{1}{\mu}\mathbf{L}_{sf_1}\mathbf{A}_{sf_{21}}\right)\mathbf{e}_{sf}, \tag{3.53}$$

where $\dot{\mathbf{e}}_{sf} = \dot{\mathbf{z}}_{sf_1} - \dot{\hat{\mathbf{z}}}_{sf_1}$ suggesting that the appropriate selection of the gain $\mathbf{L}_{sf_1}$, such that $\psi(\mathbf{A}_{sf_{11}} - \frac{1}{\mu}\mathbf{L}_{sf_1}\mathbf{A}_{sf_{21}}) < 0$, can cause the estimation error to decay to zero. Here also, the pair $(\mathbf{A}_{sf_{11}}, \mathbf{A}_{sf_{21}})$ needs to be observable, which is satisfied by Assumption 3.1. Observability of $(\mathbf{A}_{sf_{11}}, \mathbf{A}_{sf_{21}})$ in the presence of observable pair $(\mathbf{A}, \mathbf{C})$ is proved in Radisavljevic-Gajic (2015).

### 3.3.2.2 Estimation of Some States in Both Slow and Fast Modes

Consider that, $l_1$ states of $\mathbf{z}_1$ and $l_2$ states of $\mathbf{z}_2$ are not available for measurement, defined as

$$\mathbf{z}_1 = \left[\mathbf{z}_{11}^T \ \mathbf{z}_{12}^T\right]^T \text{ and } \mathbf{z}_2 = \left[\mathbf{z}_{21}^T \ \mathbf{z}_{22}^T\right]^T, \tag{3.54}$$

where $\mathbf{z}_{11} \in \Re^{n_1-l_1}, \mathbf{z}_{12} \in \Re^{l_1}, \mathbf{z}_{21} \in \Re^{l_2}$ and $\mathbf{z}_{22} \in \Re^{n_2-l_2}$. Remaining all other states, including $\mathbf{z}_{11}$, $\mathbf{z}_{22}$ and $\mathbf{z}_3$ are measurable. As a consequence, the matrices $\mathbf{A}_{ij}$ and $\mathbf{B}_i$ in (3.16) are partitioned appropriately as

$$\mathbf{A}_{11} = \begin{bmatrix} \mathbf{a}_{11} & \mathbf{a}_{12} \\ \mathbf{a}_{21} & \mathbf{a}_{22} \end{bmatrix}, \mathbf{A}_{12} = \begin{bmatrix} \mathbf{a}_{13} & \mathbf{a}_{14} \\ \mathbf{a}_{23} & \mathbf{a}_{24} \end{bmatrix}, \mathbf{A}_{13} = \begin{bmatrix} \mathbf{a}_{15} \\ \mathbf{a}_{25} \end{bmatrix}, \mathbf{A}_{21} = \begin{bmatrix} \mathbf{a}_{31} & \mathbf{a}_{32} \\ \mathbf{a}_{41} & \mathbf{a}_{42} \end{bmatrix}, \mathbf{A}_{22} = \begin{bmatrix} \mathbf{a}_{33} & \mathbf{a}_{34} \\ \mathbf{a}_{43} & \mathbf{a}_{44} \end{bmatrix},$$

$$\mathbf{A}_{23} = \begin{bmatrix} \mathbf{a}_{35} \\ \mathbf{a}_{45} \end{bmatrix}, \mathbf{A}_{31} = \begin{bmatrix} \mathbf{a}_{51}^T \\ \mathbf{a}_{52}^T \end{bmatrix}^T, \mathbf{A}_{32} = \begin{bmatrix} \mathbf{a}_{53}^T \\ \mathbf{a}_{54}^T \end{bmatrix}^T, \mathbf{A}_{33} = \mathbf{a}_{55}, \mathbf{B}_1 = \begin{bmatrix} \mathbf{b}_1 \\ \mathbf{b}_2 \end{bmatrix}, \mathbf{B}_2 = \begin{bmatrix} \mathbf{b}_3 \\ \mathbf{b}_4 \end{bmatrix}, \mathbf{B}_3 = \mathbf{b}_5,$$

where $\mathbf{a}_{ij}$ and $\mathbf{b}_i$ are of suitable dimensions. Regrouping of states as immeasurable and measurable, one can have

$$\left.\begin{aligned} \dot{\mathbf{z}}_{fs_1} &= \mathbf{A}_{fs_{11}}\mathbf{z}_{fs_1} + \mathbf{A}_{fs_{12}}\mathbf{z}_{fs_2} + \mathbf{B}_{fs_1}\mathbf{u}, \\ \dot{\mathbf{z}}_{fs_2} &= \mathbf{A}_{fs_{21}}\mathbf{z}_{fs_1} + \mathbf{A}_{fs_{22}}\mathbf{z}_{fs_2} + \mathbf{B}_{fs_2}\mathbf{u}, \\ \mathbf{y} &= \mathbf{z}_{fs_2}, \end{aligned}\right\} \tag{3.55}$$

where $\mathbf{z}_{fs_1} = \left[\mathbf{z}_{12}^T \ \mathbf{z}_{21}^T\right]^T, \mathbf{z}_{fs_2} = \left[\mathbf{z}_{11}^T \ \mathbf{z}_{22}^T \ \mathbf{z}_3^T\right]^T$ and

$$\mathbf{A}_{fs11} = \begin{bmatrix} \mathbf{a}_{22} & \mathbf{a}_{23} \\ \frac{\mathbf{a}_{32}}{\varepsilon} & \frac{\mathbf{a}_{33}}{\varepsilon} \end{bmatrix}, \mathbf{A}_{fs12} = \begin{bmatrix} \mathbf{a}_{21} & \mathbf{a}_{24} & \mathbf{a}_{25} \\ \frac{\mathbf{a}_{31}}{\varepsilon} & \frac{\mathbf{a}_{34}}{\varepsilon} & \frac{\mathbf{a}_{35}}{\varepsilon} \end{bmatrix}, \mathbf{A}_{fs21} = \begin{bmatrix} \mathbf{a}_{12} & \mathbf{a}_{13} \\ \frac{\mathbf{a}_{42}}{\varepsilon} & \frac{\mathbf{a}_{43}}{\varepsilon} \\ \frac{\mathbf{a}_{52}}{\mu} & \frac{\mathbf{a}_{53}}{\mu} \end{bmatrix}, \mathbf{A}_{fs22} = \begin{bmatrix} \mathbf{a}_{11} & \mathbf{a}_{14} & \mathbf{a}_{15} \\ \frac{\mathbf{a}_{41}}{\varepsilon} & \frac{\mathbf{a}_{44}}{\varepsilon} & \frac{\mathbf{a}_{45}}{\varepsilon} \\ \frac{\mathbf{a}_{51}}{\mu} & \frac{\mathbf{a}_{53}}{\mu} & \frac{\mathbf{a}_{55}}{\mu} \end{bmatrix},$$

$$\mathbf{B}_{fs_1} = \left[\mathbf{b}_2^T \ \frac{\mathbf{b}_3^T}{\varepsilon}\right]^T, \mathbf{B}_{fs_2} = \left[\mathbf{b}_1^T \ \frac{\mathbf{b}_4^T}{\varepsilon} \ \frac{\mathbf{b}_5^T}{\mu}\right]^T.$$

The corresponding reduced-order observer (Chen 1999) for (3.55) is

$$\dot{\hat{\mathbf{z}}}_{fs} = \mathbf{A}_{fs}\hat{\mathbf{z}}_{fs} + \mathbf{B}_{fs}\mathbf{u} + \mathbf{L}_{fs}\mathbf{y} \tag{3.56}$$

where $\mathbf{A}_{fs} = \mathbf{A}_{fs_{11}} - \mathbf{L}_{fs_1}\mathbf{A}_{fs_{21}}$, $\mathbf{B}_{fs} = \mathbf{B}_{fs_1} - \mathbf{L}_{fs_1}\mathbf{B}_{fs_2}$, $\mathbf{L}_{fs} = \mathbf{A}_{fs}\mathbf{L}_{fs_1} - \mathbf{L}_{fs_1}\mathbf{A}_{fs_2 2} + \mathbf{A}_{fs_{12}}$ with state estimation as $\hat{\mathbf{z}}_{fs} = \hat{\mathbf{z}}_{fs_1} - \mathbf{L}_{fs_1}\mathbf{z}_{fs_2}$, where $\mathbf{L}_{fs_1}$ is the reduced-order composite observer gain of order $((l_1 + l_2) \times (n_3 + (n_1 - l_1) + (n_2 - l_2)))$. This can also be derived by a two-stage design (refer Sect. 3.3.2.1). The aim of the observer (3.56) is to construct immeasurable states $\mathbf{z}_{fs_1}$ in an asymptotic manner in the sense that $\dot{\mathbf{e}}_{fs} = (\mathbf{A}_{fs_{11}} - \mathbf{L}_{fs_1}\mathbf{A}_{fs_{21}})\mathbf{e}_{fs}$ tends to zero. To achieve this goal $\mathbf{L}_{fs_1}$ is chosen such that the matrix $(\mathbf{A}_{fs_{11}} - \mathbf{L}_{fs_1}\mathbf{A}_{fs_{21}})$ is asymptotically stable. It is possible if the pair $(\mathbf{A}_{fs_{11}}, \mathbf{A}_{fs_{21}})$ is observable.

**Remark 3.1** The observer design can be made simpler for the reduced-order observer with any combination of immeasurable states by appropriately regrouping the states, as in (3.55). The three-stage design is advised if the inaccessible states contain a few states from the slow, fast, and very fast modes. If they contain states from any two modes, then a two-stage design is advised.

The novel aspects of the reduced-order observer design and their advantages are summarized here.

1. The triangular form (3.48) of (3.47) is reached by solving a non-symmetrical Riccati equation (3.23) for **P** and then the two-stage design is done by solving a Sylvester equation (3.40) for **M**, i.e., only two algebraic equations are required to be evaluated compared to three equations in Yoo and Gajic (2018), thereby reducing computation time.
2. The computational complexity and offline computation time are greatly reduced because only two transformations are performed, one to obtain the lower triangular form (3.48) and the other to obtain the block-diagonal form (3.50).
3. The system's robustness and reliability are enhanced by independently designed observer gains (in both two-stage and three-stage architectures).

## 3.4 Application to AHWR

This section corroborates the proposal of the observers, designed in Sect. 3.3, for the nonlinear system of AHWR. The AHWR model (3.15) is structured into a three-time-scale form (3.16) first and then observers are formulated for it.

### *3.4.1 Three-Time-Scale Representation of AHWR Model*

Partitioning a multi-time-scale system into several subsystems can be done based on the physical nature of the system, mathematical conditions, specific control requirements, grouping of state variables, etc. (Zhang et al. 2014). In the case of AHWR, three distinct clusters of eigenvalues help to identify the three-time-scale nature, i.e., 90 eigenvalues are grouped into 38 slow, 35 fast, and 17 very fast eigenvalues. Thus, $\varepsilon$ and $\mu$ are calculated as ratios of magnitudes of eigenvalues from respective clusters

$$\varepsilon = \frac{max|\psi_s|}{min|\psi_f|} = 5.3399 \times 10^{-1},$$
$$\mu = \frac{max|\psi_s|}{min|\psi_{ff}|} = 8.6741 \times 10^{-4}.$$

For obtaining an explicit singular perturbation form of AHWR, i.e., a model with first, second, and third subsystems with respectively slow, fast, and very fast eigenvalues and nonsingular matrices $\mathbf{A}_{22}$ and $\mathbf{A}_{33}$ (i.e., satisfying Assumption 3.3), state vector of the system (3.15) is partitioned as

$$\left.\begin{aligned} \mathbf{z}_1 &= \left[\mathbf{z}_H^T \; \mathbf{z}_X^T \; \mathbf{z}_I^T\right]^T, \\ \mathbf{z}_2 &= \left[\delta h_d \; \mathbf{z}_C^T \; \mathbf{z}_x^T\right]^T, \\ \mathbf{z}_3 &= \mathbf{z}_P. \end{aligned}\right\} \tag{3.57}$$

Accordingly, 'slow' states include regulating rod positions, xenon, and iodine concentrations, 'fast' states contain downcomer enthalpy, precursors' concentrations, and exit quality, and 'very fast' states comprise nodal powers. Recall that the system (3.15) also fulfills Assumption 3.1. Feed flow rate, $q_f$ is a function of total power and $\mathbf{B}_f$ is a column vector of dimension 90 with a single entry at the 39th position. Hence, the input $\delta q_f$ is treated separately and is not considered in the three-time-scale form of AHWR. Now, as per (3.57), submatrices $\mathbf{A}_{ij}$, $\mathbf{B}_i$ and $\mathbf{C}_i$ are obtained from (3.15) and a singularly perturbed three-time-scale structure (3.16) is obtained.

### 3.4.2 *Observer Designs*

Designing a full-order observer follows the process described in Sect. 3.3.1. This involves comparing the nonlinear AHWR system's normalized nodal powers with the observer outputs and assigning the resultant error to a composite observer gain. $\delta q_f$ and controller output $\mathbf{u}_s$ are additional inputs to an observer. The estimated states are then fed to a state feedback controller, designed in Munje et al. (2015), to generate $\mathbf{u}_s$.

After that, the method in Sect. 3.3.2.2 is used to create a reduced-order observer to estimate the concentrations of xenon, iodine, and precursors. There are two parts of the state vector (3.8), viz. measurable and immeasurable state variables, such as

$$\left.\begin{aligned} \mathbf{z}_{fs_1} &= \left[\mathbf{z}_X^T \; \mathbf{z}_I^T \; \mathbf{z}_C^T\right]^T, \\ \mathbf{z}_{fs_2} &= \left[\mathbf{z}_H^T \; \delta h_d \; \mathbf{z}_x^T \; \mathbf{z}_P^T\right]^T. \end{aligned}\right\} \tag{3.58}$$

There are 34 slow and 17 fast immeasurable states, for a total of 51 observer states. In this case, $\mathbf{u}_s$, $\delta q_f$, and normalized measurable states are given to a reduced-order observer for state estimation. To generate the control signal, the same controller (Munje et al. 2015) is fed with the estimated states and measured states.

Observer closed-loop eigenvalues are chosen six times faster than system closed-loop eigenvalues when taking into account both a quick rate of convergence for estimation error and an adequate level of noise smoothing (Radisavljevic-Gajic 2015). In addition, this ensures that the three- and two-stage designs meet Assumptions 3.5 and 3.4 and that the proper speed ratio between slow, fast, and very fast observer eigenvalues is maintained. 51 of the 90 eigenvalues of the full-order observer are carefully selected for the reduced-order observer to ensure that the eigenvalues for the concentrations of xenon, iodine, and precursors are the same in both observer designs, allowing for a reasonable comparison of observers.

### 3.4.3 Transient Simulations

The AHWR, denoted by (3.1)–(3.7), is demonstrated to attain spatial stabilization through the use of a state feedback controller (Munje et al. 2015), which is supplied by the states that are computed using suggested observers.

To demonstrate the effectiveness of recommended observers, the closed-loop response of the system to the difference in the initial conditions of estimated and actual states is first explored. Here, 0.01% perturbation is introduced in the initial conditions of the estimated states. The real reactor power and the full-order observer estimated reactor power are compared in Fig. 3.1a. It is noted that the predicted total power takes about 300 s to reach a steady-state value of 920.48 MW.[1] Nodal power estimations are displayed in Fig. 3.1b as variations in the estimation errors, calculated for azimuthal tilts (Shimjith et al. 2011). These errors also decay in about 300 s. There is also a decay in these errors in about 300 s. A plot of estimation errors for the

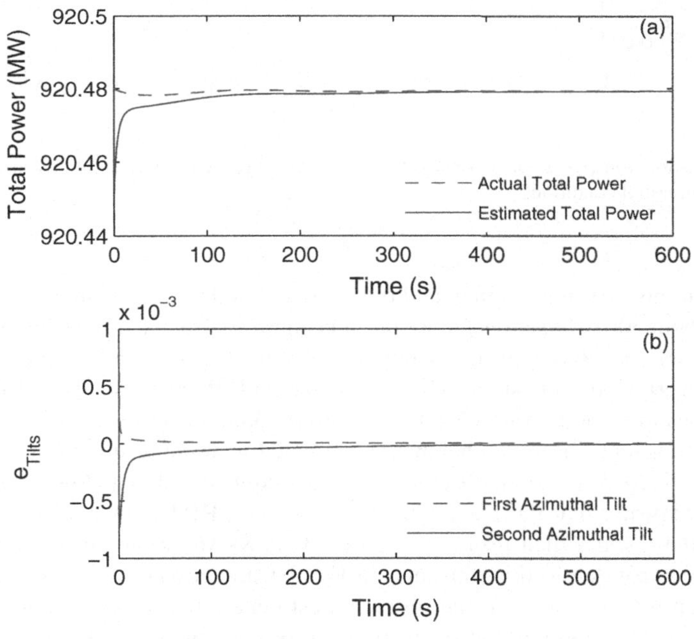


**Fig. 3.1** Results for mismatch in initial conditions with a full-order observer **a** Total reactor power and **b** Estimation errors for azimuthal tilts

[1] In this chapter, MW refers to MW (thermal).

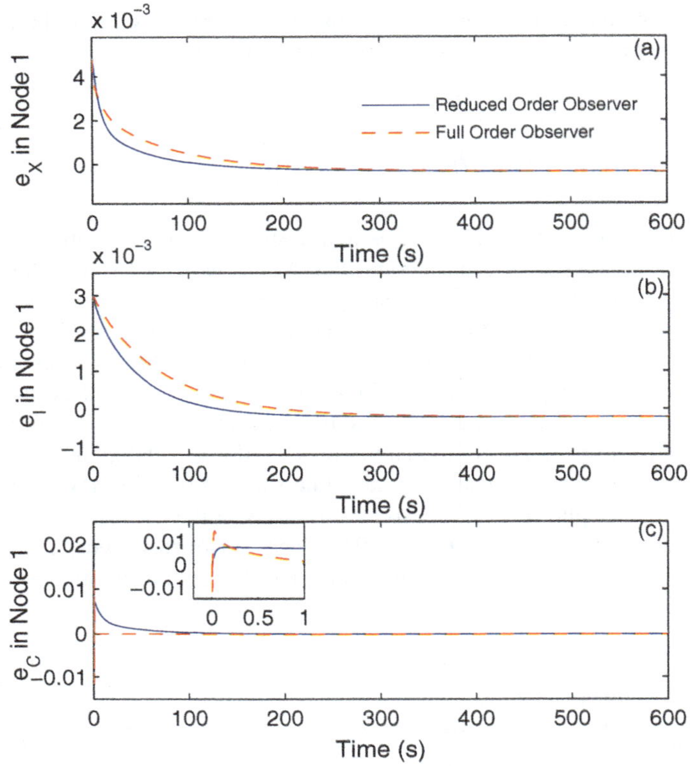


**Fig. 3.2** Estimation errors for **a** Xenon, **b** Iodine and **c** Precursor concentrations in node 1 for mismatch in initial conditions

concentrations of xenon, iodine, and precursors at node 1 is also shown in Fig. 3.2. It takes about 300 s for xenon and iodine concentrations to reach zero, but only 110 s for precursor concentrations. Even though xenon and iodine decay more slowly, their estimation errors are very small. The advised method effectively deals with the initial mismatches in all nodal variables due to a similar kind of result.

Other transients discussed below are compared with the results of the controller (Munje et al. 2015) (without observer) and fast output sampling (Munje et al. 2013).

After ensuring that the reactor is at steady state, RR4 is manually moved out by 1% after 2 s and then restored. Figure 3.3 shows variations in the total power for each control due to this transient. In Fig. 3.3, the zoomed portion (marked by a small circle) indicates that the observers estimate total power reasonably accurately. Figure 3.4 presents the estimation errors for xenon, iodine, and precursor concentrations in node 4 due to full- and reduced-order observers. It is seen that the estimation errors with a reduced-order observer converge faster than the full-order observer.

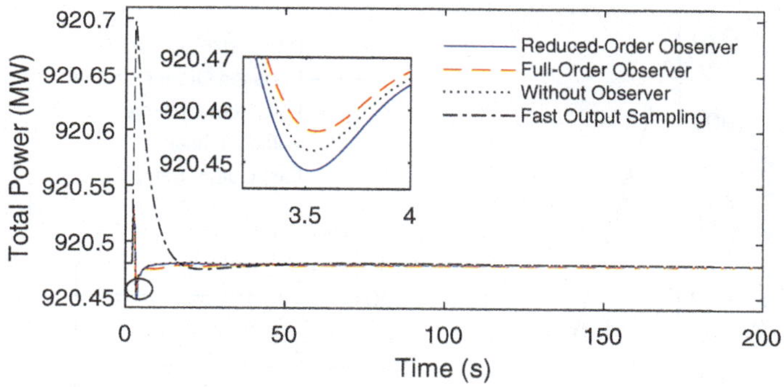


**Fig. 3.3** Results for total reactor power due to the transient generated by RR4

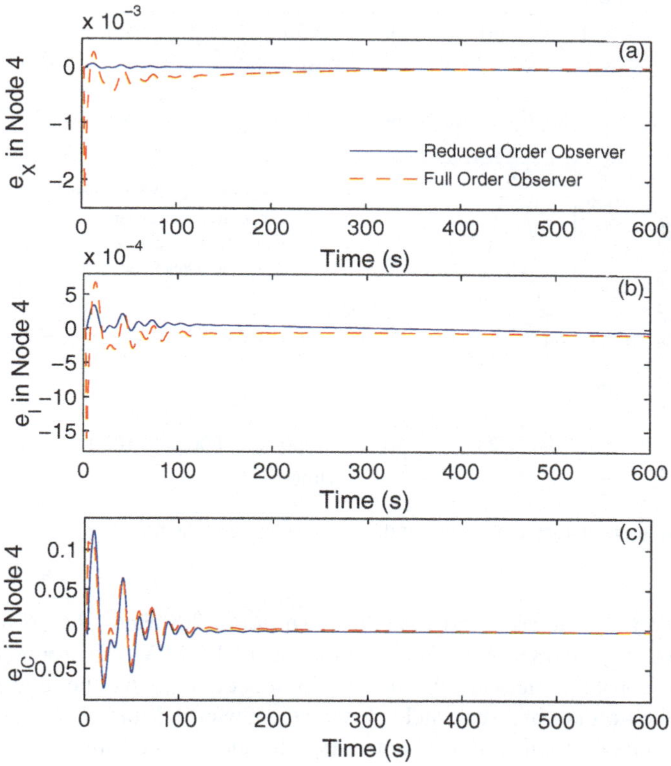


**Fig. 3.4** Estimation errors for **a** Xenon, **b** Iodine and **c** Precursor concentrations in node 4 due to the transient generated by RR4

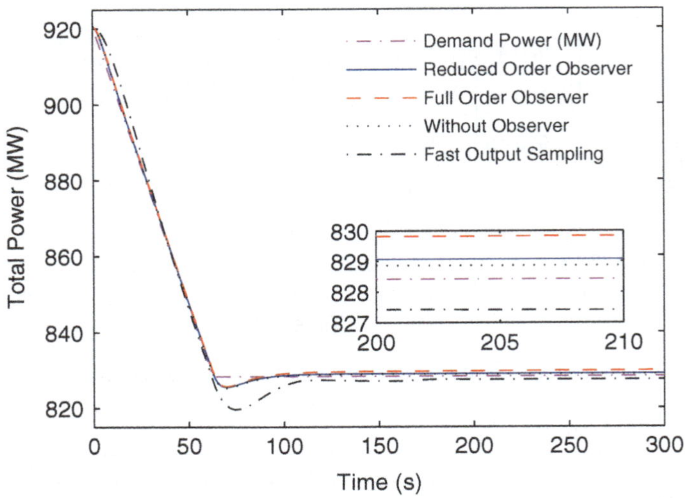


**Fig. 3.5** Total power variations during power maneuvering from 920.48 to 828.43 MW

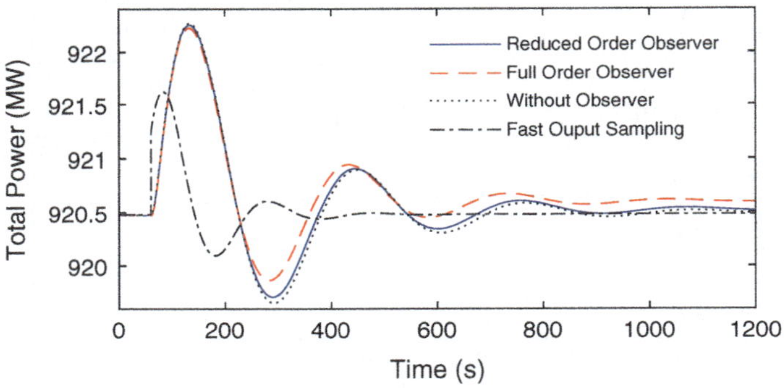


**Fig. 3.6** Variations in total reactor power due to step change in the feed flow

The power maneuvering transient is shown in Fig. 3.5. Here, the power demand is reduced from 920.48 to 828.43 MW at the rate of 1.5 MW/s, in around 61 s and is maintained constant afterward. From Fig. 3.5, it is obvious that the total power due to the reduced-order observer is held close to the power in Munje et al. (2015). Once the steady state is attained, it does not change for an extended simulation period.

Finally, to understand the behavior of the system to a disturbance, a nonlinear model is simulated with a +5% step variation in feed flow after 60 s. As a result, total power shows changes as pictured in Fig. 3.6. To keep total power at a stability level, all the RRs are moved inside the reactor by 1%. All the RRs (RR2, RR4, RR6, and RR8) under the influence of a particular controller showed similar changes (Fig. 3.7a) except for full-order observer (Fig. 3.7b). With a full-order observer, RR2 and RR6 reported the same deviations and RR6 and RR8 displayed similar variations

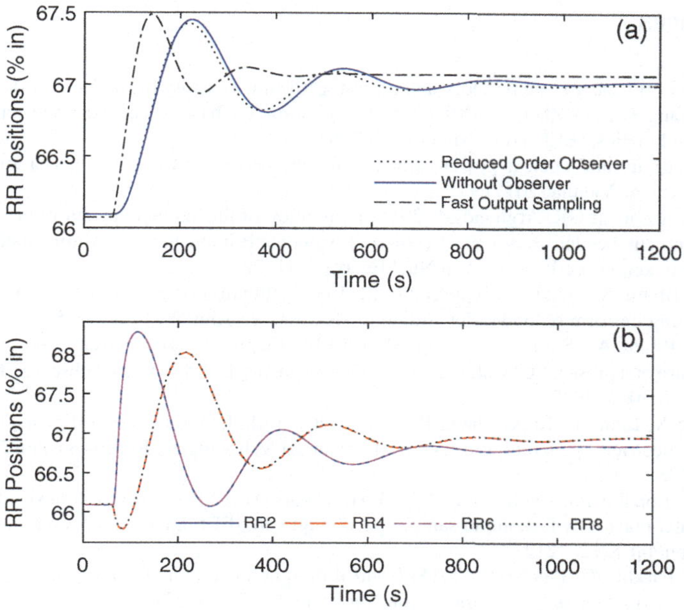


**Fig. 3.7** Variations in RR positions due to a step change in the feed flow **a** With reduced-order observer, without an observer and fast output sampling, **b** Full-order observer

as depicted in Fig. 3.7b. Even though, the performance of the fast output sampling technique is observed to be superior, for regulation and tracking its performance is not satisfactory over reduced-order observer-based control. The reduced-order observer is found better than a full-order observer in this transient too.

## 3.5 Conclusion

In this chapter, xenon, iodine, and precursor concentrations are determined by employing full- and reduced-order observers. These observers are paired with a feedback controller, and under typical transients, nonlinear system behavior is examined. To deal with higher order ill-conditioned matrices of singularly perturbed three-time scale, observers are designed by separating slow, fast, and very fast states. First, an independent three-stage design for a full-order observer is done, and then a two-stage design for a reduced-order observer. In contrast to the three-stage design, which is inventively accomplished by applying a two-stage design twice and requiring the solution of just four algebraic equations, the two-stage observer design only requires the solution of two algebraic equations. The results of the nonlinear simulation demonstrated that an AHWR system is successfully stabilized using the observers created through time-scale decomposition.

## References


Chen CT (1999) Linear system theory and design. Oxford University Press, New York

Dong Z, Huang X, Li D, Zhang Z (2016) Reactivity estimation based on an extended state observer of neutron kinetics. IEEE Trans Nucl Sci 63:2691–2697

Energy, electricity and nuclear power estimates for the period up to 2050. International Atomic Energy Agency, Vienna (2017)

Esteki MH, Ansarifar GR, Arghand M (2015) Estimation of the xenon concentration and delayed neutrons precursors densities in the pressurized-water nuclear reactors with sliding mode observer considering xenon oscillations. Ann Nucl Energy 77:1–22

Hussain S, Bhatti AI, Samee A, Qaiser SH (2013a) Estimation of precursor density of a power reactor using uniform second order sliding mode observer. Ann Nucl Energy 54:233–239

Hussain S, Bhatti AI, Samee A, Qaiser SH (2013b) Estimation of reactivity and average fuel temperature of a pressurized water reactor using sliding mode differentiator observer. IEEE Trans Nucl Sci 60:3025–3032

Li G, Wang X, Liang B, Li X, Zhang B, Zou Y (2016) Modeling and control of nuclear reactor cores for electricity generation: a review of advanced technologies. Renew Sustain Energy Rev 60:116–128

Munje RK, Londhe PS, Parkhe JG, Patre BM, Tiwari AP (2013) Spatial control of advanced heavy water reactor by fast output sampling technique. In: IEEE international conference control application, pp 1212–1217

Munje RK, Parkhe JG, Patre BM (2015) Control of xenon oscillations in advanced heavy water reactor via two-stage decomposition. Ann Nucl Energy 77:326–334

Munje RK, Patre BM, Londhe PS, Tiwari AP, Shimjith SR (2016) Investigation of spatial control strategies for AHWR: a comparative study. IEEE Trans Nucl Sci 63:1236–1246

Munje RK, Gu D, Desai R, Patre BM, Zhang W (2018) Observer-based spatial control of advanced heavy water reactor using time-scale decoupling. IEEE Trans Nucl Sci 65:2756–2766

Naidu DS (1988) Singular perturbation methodology in control systems. Peter Peregrinus Ltd, London U.K

O'Reilly J (1980) Dynamical feedback control for a class of singularly perturbed linear systems using a full-order observer. Int J Control 31:1–10

Park YH, Cho NZ (1992) Estimation of neutron flux and xenon distributions via observer-based control theory. Nucl Sci Eng 111:66–81

Patel SB, Mukhopadhyay S, Tiwari AP (2018) Estimation of reactivity and delayed neutron precursors' concentrations using a multiscale extended Kalman filter. Ann Nucl Energy 111:666–675

Radisavljevic-Gajic V (2015) Full- and reduced-order linear observer implementations in Matlab/Simulink [Lecture Notes]. IEEE Control Syst 35:91–101

Radisavljevic-Gajic V, Rose P (2014) A new two-stage design of feedback controllers for a hydrogen gas reformer. Int J Hydrog Energy 39:11738–11748

Radisavljevic-Gajic V, Milanovic M, Clayton G (2017) Three-stage feedback controller design with application to three-time-scale linear control system. ASME J Dyn Syst, Meas, Control 139:1–10

Shimjith SR, Tiwari AP, Bandyopadhyay B, Patil RK (2011) Spatial stabilization of advanced heavy water reactor. Ann Nucl Energy 38:1545–1558

Shouse KR, Taylor DG (1993) Reduced-order discrete-time recursive observers for linear continuous-time singularly perturbed systems. In: Proceedings of the American control conference, pp 2128–2129

Sinha RK, Kakodkar A (2006) Design and development of the AHWR-the Indian thorium fuelled innovative nuclear reactor. Nucl Eng Des 236:683–700

Tiwari AP, Bandyopadhyay B (1998) Control of xenon induced spatial oscillations in a large PHWR. In: Proceedings of IEEE international conference on global connectivity in energy, computer, communication and control, vol 1, pp 178–181

Tsui CC (2015) Observer design—a survey. Int J Autom Comput 12:50–61

Wang P, Aldemir T, Utkin VI (2001) Estimation of xenon concentration and reactivity in nuclear reactors using sliding mode observers. In: Proceedings IEEE conference decision and control, vol 2, pp 1801–1806
Yang C, Zhang L, Zhou L (2016) Observer design for singularly perturbed systems with multirate sampled and delayed measurements. J Dyn Syst, Meas, Control 138:051007–051009
Yeganeh MHZ, Ansarifar GR (2018) Estimation of the poisons reactivity in the P.W.R nuclear reactors using modified higher order sliding mode observer based on the multi-point nuclear reactor model. Ann Nucl Energy 112:158–169
Yoo H, Gajic Z (2018) New designs of linear observers and observer-based controllers for singularly perturbed linear systems. IEEE Trans Auto Control 63:3904–3911
Yoo H, Gajic Z (2017) New designs of reduced order observer-based controllers for singularly perturbed linear systems. Math Probl Eng 1–14
Zhang Y, Naidu DS, Cai CX, Zou Y (2014) Singular perturbations and time scales in control theories and applications: an overview 2002–2012. Int J Info Syst Sci 9:1–36

# Chapter 4
# Observer-Based Output Feedback Integral Control for Coal-Fired Power Plant

**Abstract** This chapter explores the control of multivariable coal-fired power plants (CFPP) characterized by interconnected nonlinear subsystems with varying time lags. Employing singular perturbation and time-scale theories, the study develops an integral state feedback controller with an observer using a three-time-scale approach. This controller ensures zero steady-state error, utilizes a three-stage feedback control with a two-time-scale separation for reduced complexity, and employs an observer constructed through time-scale decoupling. Practical testing on a 300 MW CFPP model demonstrates superior performance compared to another published controller, highlighting the effectiveness of the proposed three-time-scale control strategy in managing plant behavior under various transients.

**Keywords** Integral control · Power plant control · Three-stage design · Three-time-scale system · Singular perturbation

## 4.1 Introduction

It is projected that global energy consumption will rise by 28% by 2040. Fossil fuels continue to lead the globe in energy production, accounting for 77% of global energy consumption in 2040, even if the steady rise of renewable energy sources is projected to reach 2.3% per year from 2015 to 2040 (Annual Energy Outlook 2017). Additionally, it is anticipated that until 2040, China's coal-fired power output, which presently accounts for roughly 70.31% of the country's total generation (Key World Energy Statistics 2017), would largely remain stable, with renewable energy sources making up the difference as energy consumption rises. Nevertheless, as renewable energy sources are integrated into the grid, their unpredictability and fluctuation will make operating and controlling conventional coal-fired power plants (CFPP) more difficult. Therefore, for CFPP, an effective and trustworthy control method is needed.

Over the years, many models and controls have been investigated and suggested (Chandrasekharan et al. 2014; Wu et al. 2015), such as, generalized predictive control (Peng et al. 2002) and coordinated control (Li et al. 2005; Liu et al. 2010), etc. In

 
R. Munje et al., *Multi-stage Feedback Control Design for Multi-time-scale System*, Springer Tracts in Electrical and Electronics Engineering,
https://doi.org/10.1007/978-981-97-4592-0_4

Alamoodi and Daoutidis (2016), a nonlinear control for the CFPP that uses feedback linearization in conjunction with dead-time compensation to produce steady closed-loop and effective tracking performance is reported. An extended model containing the dynamics of superheated steam temperature is also treated using the same methods (Alamoodi and Daoutidis 2017). While Alamoodi and Daoutidis (2016, 2017) yields excellent results, the controller designs and models under study are difficult to configure in a traditional CFPP. In a separate study, Sun et al. (2016) developed a simple model based on the direct energy balance (DEB) and used gain scheduling PI control to meet the tracking requirements. However, it is observed that this controller is ineffective at managing frequent changes in coal quality. Consequently, an active disturbance rejection control is developed in Sun et al. (2017). On the other hand, it neutralizes the disruption and ambiguity quite gradually. An optimal control technique for a boiler-turbine model to effectively manage uncertainty is described in Moradi and Vossoughi (2016). Nevertheless, it is an observer-dependent method relying on state feedback. Thus, in Gamal et al. (2016), an observer-based optimum linear quadratic regulator (LQR) is developed and tested on the model given in Moradi and Vossoughi (2016). Similarly, $H_\infty$-LQR (Wei and Fang 2017) is also effectively tested for boosting demand power variation in a variety of operations on a big boiler-turbine unit.

As a result of high dimensionality and model rigidity, coal-fired power plants frequently display a multi-time-scale characteristic. However, for such an ill-conditioned system of the power plant, the direct design of feedback controls (Moradi and Vossoughi 2016; Gamal et al. 2016; Wei and Fang 2017) leads to higher gain values, which makes it challenging for real-time realization. Time-scale approaches and singular perturbation can be used to solve this issue (Zhang et al. 2014). Examples of time-scale theory applied to power plants can be found in Naidu and Rao (1985); Patre et al. (1999); Cori and Maffezzoni (2016). The time-scale structure of the power plant is discussed in Naidu and Rao (1985). A periodic output feedback controller is designed in Patre et al. (1999) using a two-time-scale method. A two-time-scale quasi-steady-state approach is examined in Cori and Maffezzoni (2016) for the ninth-order power plant. The quasi-steady-state approach is approximate, due to which the goal of precisely positioning subsystem eigenvalues is not satisfied fully. Two-stage designs can be implemented as a result (Phillips 1980; Radisavljevic-Gajic and Rose 2014; Yoo and Gajic 2018). In actuality, the three-time-scale approach offers more control design options than the two-time-scale method. Studies in Naidu and Ravinder (1985); Singh et al. (1986) present the block-diagonalization of the three-time-scale system into ‘slow’, ‘quick’, and ‘very fast’ modes. However, control design aspects are not covered in Naidu and Ravinder (1985) and Singh et al. (1986). In the recent past, Radisavljevic-Gajic et al. (2017) proposed a reduced computation three-stage feedback control architecture for a three-time-scale system.

This chapter proposes and investigates an integral control for a three-time-scale system for the CFPP with an observer (Munje et al. 2020). Since the usual CFPP control aims are disturbance rejection and excellent set-point tracking, state feedback with integrated action is advised for a three-time-scale system. In contrast to Radisavljevic-Gajic et al. (2017), a simplified three-stage design provided in Chap. 2

elicits this feedback control and prevents system ill-conditioning. Additionally, an inequality between the suggested approach and the method of Radisavljevic-Gajic et al. (2017) is derived, and when satisfied, it leads to a significant reduction in online computation time. Next, this controller is used in the real-world 300 MW[1] coal-fired power plant, where the full-order observer—which is likewise designed using a three-stage process—is used to estimate the states. The difference between the three-stage feedback control design in Chap. 2 and this chapter is the explicit consideration of singular perturbation parameters in the design, which is not considered in Chap. 2.

## 4.2 CFPP Model Description

In Sun et al. (2016), direct energy balance control of a nonlinear 300 MW CFPP model, situated in Guangdong Province, China is discussed. The detailed model is given by

$$\frac{dq_f}{dt} = \frac{1}{c_0}\left[u_B(t-\tau) - q_f\right], \tag{4.1}$$

$$\frac{dD_b}{dt} = \frac{1}{c_5}\left[2.46k_c q_f^{1.230} - D_b\right], \tag{4.2}$$

$$\frac{dp_b}{dt} = \frac{1}{c_6}\left[D_b - 42.51 p_b^{0.956}\sqrt{p_b - p_T}\right], \tag{4.3}$$

$$\frac{dp_T}{dt} = \frac{1}{c_7}\left[42.51 p_b^{0.956}\sqrt{p_b - p_T} - D_T\right], \tag{4.4}$$

$$\frac{dp_1}{dt} = \frac{1}{c_1}\left[0.0083\mu_T p_T - p_1\right], \tag{4.5}$$

$$\frac{dD_T}{dt} = \frac{1}{c_2}\left[74.74 p_1 - D_T\right], \tag{4.6}$$

with output equations

$$N_e = 0.86 D_T^{0.852}, \tag{4.7}$$

$$Q_m = p_1 + \frac{C_b}{c_6}\left[D_b - 42.51 p_b^{0.956}\sqrt{p_b - p_T}\right], \tag{4.8}$$

where $c_0, c_1, c_2, c_5, c_6$ and $c_7$ are inertia constants, $\tau$ is a time delay, $k_c$ is a normalized coefficient denoting the influence of coal quality and $C_b$ is thermal storage coefficient. The states $q_f$, $D_b$, $p_b$, $p_T$, $p_1$ and $D_T$ form the state vector $\mathbf{z}$. In this, $q_f$ is the coal mass flow rate in-flowing into the furnace (t/h), $D_b$ is production rate of steam (t/h), $p_b$ is pressure of drum (MPa), $p_T$ is live steam pressure (MPa), $p_1$ is pressure in governing stage (MPa) and $D_T$ is mass flow rate of inlet steam (t/h). Further, $N_e$ and

[1] In this chapter, MW refers to MW (electrical).

$Q_m$ are the outputs, respectively, electrical power (MW) and heat provided by the boiler (MPa) and $u_B$ and $\mu_T$ are control inputs, i.e., boiler demand (t/h) and throttle opening position (%) respectively. Following are the actuator limit constraints for inputs:

$$0 \leq u_B \leq 150 \text{ and } 0 \leq \frac{du_B}{dt} \leq 0.1,$$
$$0 \leq \mu_T \leq 100 \text{ and } 0 \leq \frac{d\mu_T}{dt} \leq 0.1.$$

The system (4.1)–(4.8) is linearized around steady-state with $N_e = 285.9$ MW, $Q_m = 12.19$ MPa, $u_B = 122.6$ t/h and $\mu_T = 93.5\%$. The inertia constants, $\tau$ and $C_b$ are given in Sun et al. (2016) and $k_c$ is taken as 100% under steady-state (Sun et al. 2017). The linear equations, so obtained, are expressed in standard state-space form as

$$\dot{\mathbf{z}} = \mathbf{Az} + \mathbf{Bu}, \tag{4.9}$$
$$\mathbf{y} = \mathbf{Cz}, \tag{4.10}$$

where vectors **z**, **u** and **y** are respectively

$$\mathbf{z} = \begin{bmatrix} q_f & D_b & p_b & p_T & p_1 & D_T \end{bmatrix}^T, \tag{4.11}$$
$$\mathbf{u} = \begin{bmatrix} u_B(t-\tau) & \mu_T \end{bmatrix}^T, \quad \mathbf{y} = \begin{bmatrix} N_e & Q_m \end{bmatrix}^T. \tag{4.12}$$

Note that the time delay $\tau$ in the input $u_B$ is simulated as transport delay $e^{-\tau s}$ (Sun et al. 2016; Zhang 2012). The open loop system (4.9) is observed to be stable, i.e., $\psi(\mathbf{A}) < 0$, where $\psi(.)$ indicates eigenvalues. There are three groups of the eigenvalues for this system, i.e., (−0.0026, −0.0071), as 'slow' eigenvalues, (−0.0454) as 'fast' eigenvalue and ($-0.1306 \pm j0.0361$, −0.2497) as 'very fast' eigenvalues. The ratios of the 'slow' versus 'fast' versus 'very fast' modes are $\varepsilon = 0.1564$ and $\mu = 0.0524$ respectively. The above system is also ensured to be controllable and observable. Control of a multi-input-multi-output CFPP, presented here, is challenging because of the simultaneous existence of system states with varying response times and nonlinear system dynamics. The power plant's modules for pressure-power, steam pressure, and coal steam are highly integrated. The nonlinearity is brought on by this. Conversely, the system becomes numerically ill-conditioned with strong couplings between states with different time speeds. Because of this, power plants require more advanced control systems for both their narrow and wide-ranging operations. Therefore, a well-designed controller is needed for the power plant to ensure optimum efficiency and safety.

The primary control goals are as follows: (1) The throttle pressure reverse change should occur within ±0.4 MPa and the power tracking rate should be 1.5–2% of the total load per minute (Sun et al. 2016). (2) The output and state variables need to be quickly regulated, regardless of any transients in the power plant's input variables. (3) The plant's output variables have to be insensitive to significant parameter fluctuations and coal quality.

## 4.3 Control Design Method

This section describes the design and implementation of state feedback with integral control for a three-time-scale system using an observer. An overview of integral state feedback control with an observer is initially given. A proposal is then made to apply it to a three-time-scale system.

### 4.3.1 Observer-Based Output Feedback Integral Control

Consider a linear time-invariant continuous-time system as

$$\dot{\mathbf{z}} = \mathbf{Az} + \mathbf{Bu}, \tag{4.13}$$

$$\mathbf{y} = \mathbf{Cz}, \tag{4.14}$$

where $\mathbf{z} \in \Re^n$, $\mathbf{u} \in \Re^m$ and $\mathbf{y} \in \Re^p$. The matrices $\mathbf{A}$, $\mathbf{B}$ and $\mathbf{C}$ are of suitable dimensions.

**Assumption 4.1** $(\mathbf{A}, \mathbf{B}, \mathbf{C})$ is controllable and observable.

**Assumption 4.2** The system $\mathbf{G}(s) = \mathbf{C}[s\mathbf{I}_n - \mathbf{A}]^{-1}\mathbf{B}$ does not have transmission zeros at the origin (Chen 1999).

Designing feedback and observer gains requires adherence to Assumption 4.1, whereas tracking control that encourages integral action is required to meet Assumption 4.2. Achieving asymptotic tracking may be impossible if the parameters of the system (4.13)–(4.14) are not precisely determined. Here, it could be accomplished using an integral control. In addition, if the system is stable when it is designed, the closed-loop system with integral action becomes more robust. For the same, let us consider the system (4.13)–(4.14) with

$$\dot{\mathbf{z}}_y = \mathbf{y} - \mathbf{y}_{ref}, \tag{4.15}$$

where $\mathbf{y}_{ref} \in \Re^p$ is reference output vector. Thus, from (4.13)–(4.15), the overall system is assembled as

$$\dot{\mathbf{z}}_{aug} = \hat{\mathbf{A}}\mathbf{z}_{aug} + \hat{\mathbf{B}}\mathbf{u} + \hat{\mathbf{E}}\mathbf{y}_{ref}, \tag{4.16}$$

$$\mathbf{y} = \hat{\mathbf{C}}\mathbf{z}_{aug}, \tag{4.17}$$

where $\mathbf{z}_{aug} = \left[\mathbf{z}_y^T \ \mathbf{z}^T\right]^T$ and

$$\hat{\mathbf{A}} = \begin{bmatrix} \mathbf{0} & \mathbf{C} \\ \mathbf{0} & \mathbf{A} \end{bmatrix}, \hat{\mathbf{B}} = \begin{bmatrix} \mathbf{0} \\ \mathbf{B} \end{bmatrix}, \hat{\mathbf{E}} = \begin{bmatrix} -\mathbf{I}_p \\ \mathbf{0} \end{bmatrix}, \hat{\mathbf{C}} = \begin{bmatrix} \mathbf{0} \\ \mathbf{C}^T \end{bmatrix}^T,$$

in which $\mathbf{I}_p$ is an identity matrix of order $p$ and $\mathbf{0}$ is null matrix of proper order. The controllability of (4.16) is guaranteed, if Assumptions 4.1–4.2 are satisfied (Chen 1999). Thus, all the eigenvalues of (4.16) can be placed arbitrarily by selecting feedback gain $\hat{\mathbf{K}}$. As a result, the control input

$$\mathbf{u} = -\hat{\mathbf{K}}\mathbf{z}_a ug = -\left[\mathbf{K}_y \ \mathbf{K}\right]\left[\mathbf{z}_y^T \ \mathbf{z}^T\right]^T, \tag{4.18}$$

is applied to the system (4.16). Then

$$\dot{\mathbf{z}}_{aug} = \mathbf{A}_{cl}\mathbf{z}_{aug} + \hat{\mathbf{E}}\mathbf{y}_{ref}, \tag{4.19}$$

$$\mathbf{y} = \hat{\mathbf{C}}\mathbf{z}_{aug}, \tag{4.20}$$

where $\mathbf{A}_{cl} = \hat{\mathbf{A}} - \hat{\mathbf{B}}\hat{\mathbf{K}}$. An asymptotic set-point tracking can also be interpreted by the transfer function of (4.19)–(4.20) as

$$\mathbf{G}_{cl}(s) = \hat{\mathbf{C}}[s\mathbf{I}_{n+p} - \mathbf{A}_{cl}]^{-1}\hat{\mathbf{E}}. \tag{4.21}$$

Here, the DC-gain of (4.21) is 1, implying reference tracking. However, most of the time system states are unavailable for feedback. In that case, using Assumption 4.1, a state observer can be built. So the control (4.18) is replaced with

$$\mathbf{u} = -\mathbf{K}_y\mathbf{z}_y - \mathbf{K}\hat{\mathbf{z}}, \tag{4.22}$$

where $\hat{\mathbf{z}}$ is an estimated state vector with $\mathbf{e} = \mathbf{z} - \hat{\mathbf{z}}$ satisfying

$$\dot{\mathbf{e}} = (\mathbf{A} - \mathbf{LC})\mathbf{e}, \tag{4.23}$$

with $\mathbf{L}$ as an observer gain for (4.13)–(4.14). Then control (4.22)

$$\mathbf{u} = -\mathbf{K}_y\mathbf{z}_y - \mathbf{K}\mathbf{z} + \mathbf{K}\mathbf{e}. \tag{4.24}$$

Substituting (4.24) in (4.16) and combining with (4.23) yield

$$\dot{\zeta} = \mathbf{A}_{aug}\zeta + \mathbf{B}_{aug}\mathbf{y}_{ref}, \tag{4.25}$$

$$\mathbf{y} = \mathbf{C}_{aug}\zeta, \tag{4.26}$$

where $\zeta = \left[\mathbf{z}_y^T \ \mathbf{z}^T \middle| \mathbf{e}^T\right]^T$ and

$$\mathbf{A}_{aug} = \left[\begin{array}{cc|c} \mathbf{0} & \mathbf{C} & \mathbf{0} \\ -\mathbf{B}\mathbf{K}_y & \mathbf{A} - \mathbf{B}\mathbf{K} & \mathbf{B}\mathbf{K} \\ \hline \mathbf{0} & \mathbf{0} & \mathbf{A} - \mathbf{L}\mathbf{C} \end{array}\right] = \begin{bmatrix} \hat{\mathbf{A}} - \hat{\mathbf{B}}\hat{\mathbf{K}} & \hat{\mathbf{B}}\mathbf{K} \\ \mathbf{0} & \mathbf{A} - \mathbf{L}\mathbf{C} \end{bmatrix},$$

$$\mathbf{B}_{aug} = \left[\begin{array}{c} -\mathbf{I}_p \\ \mathbf{0} \\ \hline \mathbf{0} \end{array}\right] = \begin{bmatrix} \hat{\mathbf{E}} \\ \mathbf{0} \end{bmatrix}, \ \mathbf{C}_{aug} = \left[\begin{array}{c} \mathbf{0} \\ \mathbf{C}^T \\ \hline \mathbf{0} \end{array}\right]^T = \begin{bmatrix} \hat{\mathbf{C}}^T \\ \mathbf{0} \end{bmatrix}^T.$$

It is important to note that

$$\mathbf{G}_{cl}(s) = \mathbf{C}_{aug}[s\mathbf{I}_{2n+p} - \mathbf{A}_{aug}]^{-1}\mathbf{B}_{aug}, \tag{4.27}$$

is similar to (4.21). Thus, the reference tracking properties of the system are conserved in the presence of plant parameter variations and disturbances whether the integral state feedback or observer-based integral output feedback is employed.

### 4.3.2 Controller Synthesis for Three-Time-Scale System

Before proceeding, the following Theorem 4.1 is introduced.

**Theorem 4.1** *If the system (4.13)–(4.14) exhibits three-time-scale property, with $n_1$ 'slow', $n_2$ 'fast' and $n_3$ 'very fast' modes, such that $n_1 + n_2 + n_3 = n$, then the system (4.16)–(4.17) of order $n + p$ also bears three-time-scale form with 'slow', 'fast' and 'very fast' modes, respectively of orders $n_1 + p$, $n_2$ and $n_3$.*

***Proof*** The eigenvalues of the systems (4.13)–(4.14) and (4.16)–(4.17) can respectively be given by the characteristic equations

$$D_1(s) = \det\left[s\mathbf{I}_n - \mathbf{A}\right], \tag{4.28}$$

$$D_2(s) = \det\begin{bmatrix} s\mathbf{I}_p & -\mathbf{C} \\ \mathbf{0} & s\mathbf{I}_n - \mathbf{A} \end{bmatrix}. \tag{4.29}$$

The solution of (4.29) appears to add $p$ eigenvalues at the origin to the solution of (4.28), which already contains eigenvalue modes for $n_1$ (slow), $n_2$ (fast), and $n_3$ (very fast). Absolute eigenvalues determine the system's modes; eigenvalues close to the imaginary axis are grouped as 'slow', while those farther and closer to the imaginary axis are grouped as 'fast' and 'very fast', respectively. Since the eigenvalues in the 'fast' and 'very fast' modes remain unchanged, the $p$ eigenvalues at origin are counted in the 'slow' mode, raising the order of the slow mode of the system (4.16)–(4.17) from $n_1$ to $n_1 + p$. □

The majority of 'fast' and 'very fast' modes are stable, while 'slow' modes are not. Additionally, while trying to manage the integral state feedback, more '$p$' poles are added at the origin, which increases instability even more. In such systems, the simple design in (4.18) is computationally difficult. This is also true for observer gain in (4.23). Therefore, by regrouping states, these gains are designed in three steps using the two-time-scale technique. Assumption 4.1 also holds for decoupled subsystems since the control architecture is proposed on similarity transformations. Here, the design procedure for $\hat{\mathbf{K}}$ is conveyed first and then for $\mathbf{L}$.

#### 4.3.2.1 Design of Feedback Gain ($\hat{\mathbf{K}}$)

Let (4.13)–(4.14) be represented into standard singularly perturbed three-time-scale form

$$\begin{bmatrix} \dot{\mathbf{z}}_{o1} \\ \dot{\mathbf{z}}_{o2} \\ \dot{\mathbf{z}}_{o3} \end{bmatrix} = \begin{bmatrix} \mathbf{A}_{o11} & \mathbf{A}_{o12} & \mathbf{A}_{o13} \\ \frac{\mathbf{A}_{o21}}{\varepsilon} & \frac{\mathbf{A}_{o22}}{\varepsilon} & \frac{\mathbf{A}_{o23}}{\varepsilon} \\ \frac{\mathbf{A}_{o31}}{\mu} & \frac{\mathbf{A}_{o32}}{\mu} & \frac{\mathbf{A}_{o33}}{\mu} \end{bmatrix} \begin{bmatrix} \mathbf{z}_{o1} \\ \mathbf{z}_{o2} \\ \mathbf{z}_{o3} \end{bmatrix} + \begin{bmatrix} \mathbf{B}_{o1} \\ \frac{\mathbf{B}_{o2}}{\varepsilon} \\ \frac{\mathbf{B}_{o3}}{\mu} \end{bmatrix} \mathbf{u}, \tag{4.30}$$

$$\mathbf{y} = \begin{bmatrix} \mathbf{C}_{o1} & \mathbf{C}_{o2} & \mathbf{C}_{o3} \end{bmatrix} \begin{bmatrix} \mathbf{z}_{o1}^T & \mathbf{z}_{o2}^T & \mathbf{z}_{o3}^T \end{bmatrix}^T, \tag{4.31}$$

where $\mathbf{z}_{o1} \in \Re^{n_1}$, $\mathbf{z}_{o2} \in \Re^{n_2}$, $\mathbf{z}_{o3} \in \Re^{n_3}$ are respectively slow, fast and very fast states, $\mathbf{u} \in \Re^m$ is input and $\mathbf{y} \in \Re^p$ is output. $\mathbf{A}_{oij}$, $\mathbf{B}_{oi}$ and $\mathbf{C}_{oi}$ are of appropriate dimensions (subscript '$o$' stands for original system). Parameters $\varepsilon$ and $\mu$ are speed ratios of $\mathbf{z}_{o1}$ versus $\mathbf{z}_{o2}$ versus $\mathbf{z}_{o3}$ states, such that $\mu \ll \varepsilon \ll 1$.

**Assumption 4.3** $\mathbf{A}_{o22}$ and $\mathbf{A}_{o33}$ are invertible (Naidu 1988).

This is a standard assumption, needed to get solutions of algebraic equations treated in this subsection. Now, augmenting the system (4.30)–(4.31) with (4.15), following singularly perturbed form of (4.16)–(4.17), as proved in Theorem 4.1, is obtained

$$\begin{bmatrix} \dot{\mathbf{z}}_1 \\ \dot{\mathbf{z}}_2 \\ \dot{\mathbf{z}}_3 \end{bmatrix} = \begin{bmatrix} \mathbf{A}_{11} & \mathbf{A}_{12} & \mathbf{A}_{13} \\ \frac{\mathbf{A}_{21}}{\varepsilon} & \frac{\mathbf{A}_{22}}{\varepsilon} & \frac{\mathbf{A}_{23}}{\varepsilon} \\ \frac{\mathbf{A}_{31}}{\mu} & \frac{\mathbf{A}_{32}}{\mu} & \frac{\mathbf{A}_{33}}{\mu} \end{bmatrix} \begin{bmatrix} \mathbf{z}_1 \\ \mathbf{z}_2 \\ \mathbf{z}_3 \end{bmatrix} + \begin{bmatrix} \mathbf{B}_1 \\ \frac{\mathbf{B}_2}{\varepsilon} \\ \frac{\mathbf{B}_3}{\mu} \end{bmatrix} \mathbf{u} + \begin{bmatrix} \hat{\mathbf{E}} \\ \mathbf{0} \\ \mathbf{0} \end{bmatrix} \mathbf{y}_{ref}, \tag{4.32}$$

$$\mathbf{y} = \begin{bmatrix} \mathbf{C}_1 & \mathbf{C}_2 & \mathbf{C}_3 \end{bmatrix} \begin{bmatrix} \mathbf{z}_1^T & \mathbf{z}_2^T & \mathbf{z}_3^T \end{bmatrix}^T, \tag{4.33}$$

where $\mathbf{z}_1 = \begin{bmatrix} \mathbf{z}_y^T & \mathbf{z}_{o1}^T \end{bmatrix}^T$, $\mathbf{z}_2 = \mathbf{z}_{o2}$ and $\mathbf{z}_3 = \mathbf{z}_{o3}$ with appropriate regrouping of matrices $\mathbf{A}_{ij}$, $\mathbf{B}_i$ and $\mathbf{C}_i$. Matrix $\hat{\mathbf{E}} = \begin{bmatrix} -\mathbf{I}_p & \mathbf{0} \end{bmatrix}^T$ with $\mathbf{0}$ as null matrix of dimension $n_1 \times p$. As assumed that the matrices $\mathbf{A}_{o22}$ and $\mathbf{A}_{o33}$ are invertible, $\mathbf{A}_{22}$ and $\mathbf{A}_{33}$ are also invertible. Because, $\mathbf{A}_{22} = \mathbf{A}_{o22}$ and $\mathbf{A}_{33} = \mathbf{A}_{o33}$. Here, the main objective is to design a feedback gain $\hat{\mathbf{K}}$, so that the system (4.16)–(4.17) (i.e., (4.32)–(4.33)) is stabilized, i.e., $\psi(\hat{\mathbf{A}} - \hat{\mathbf{B}}\hat{\mathbf{K}}) < 0$. As the application of input (4.18) to the system (4.16) keeps the term $\hat{\mathbf{E}}\mathbf{y}_{ref}$ in (4.19) unchanged, it is not considered further in designing control.

First of all, system (4.32) is transformed into a lower triangular form and then a three-stage design is carried out. For simplifying the design, consider state regrouping of the system (4.32) as

$$\mathbf{z}_a = \begin{bmatrix} \mathbf{z}_1^T & \mathbf{z}_2^T \end{bmatrix}^T \text{ and } \mathbf{z}_b = \mathbf{z}_3. \tag{4.34}$$

Therefore, the system (4.32) can be rewritten as

$$\begin{bmatrix} \dot{\mathbf{z}}_a \\ \dot{\mathbf{z}}_b \end{bmatrix} = \begin{bmatrix} \bar{\mathbf{A}}_{11} & \bar{\mathbf{A}}_{12} \\ \frac{\bar{\mathbf{A}}_{21}}{\mu} & \frac{\bar{\mathbf{A}}_{22}}{\mu} \end{bmatrix} \begin{bmatrix} \mathbf{z}_a \\ \mathbf{z}_b \end{bmatrix} + \begin{bmatrix} \bar{\mathbf{B}}_1 \\ \frac{\bar{\mathbf{B}}_2}{\mu} \end{bmatrix} \mathbf{u}, \tag{4.35}$$

where $\bar{\mathbf{A}}_{ij}$ and $\bar{\mathbf{B}}_i$ are reorganized accordingly. The system (4.35) is the two-time-scale representation of the system (4.32), where $n_1 + p$ 'slow' and $n_2$ 'fast' states are merged to form $(n_1 + n_2 + p)$ 'slow' states and $n_3$ 'very fast' states are taken as 'fast' states. Now, the application of transformation

$$\mathbf{z}_s = \mathbf{z}_a + \mu \mathbf{P} \mathbf{z}_b, \tag{4.36}$$

to the system (4.35), produces

$$\begin{bmatrix} \dot{\mathbf{z}}_s \\ \dot{\mathbf{z}}_b \end{bmatrix} = \begin{bmatrix} \bar{\mathbf{A}}_s & \mathbf{0} \\ \frac{\bar{\mathbf{A}}_{21}}{\mu} & \frac{\bar{\mathbf{A}}_f}{\mu} \end{bmatrix} \begin{bmatrix} \mathbf{z}_s \\ \mathbf{z}_b \end{bmatrix} + \begin{bmatrix} \bar{\mathbf{B}}_s \\ \frac{\bar{\mathbf{B}}_2}{\mu} \end{bmatrix} \mathbf{u}, \tag{4.37}$$

where $\bar{\mathbf{A}}_s = \bar{\mathbf{A}}_{11} + \mathbf{P}\bar{\mathbf{A}}_{21}, \bar{\mathbf{A}}_f = \bar{\mathbf{A}}_{22} - \mu \bar{\mathbf{A}}_{21} \mathbf{P}$ and $\bar{\mathbf{B}}_s = \bar{\mathbf{B}}_1 + \mathbf{P}\bar{\mathbf{B}}_2$. As $\mathbf{A}_{33}$, i.e., $\bar{\mathbf{A}}_{22}$ is invertible (Assumption 4.3), the unique solution of $\mathbf{P}$ can be obtained by solving

$$\bar{\mathbf{A}}_{12} + \mathbf{P}\bar{\mathbf{A}}_{22} - \mu(\bar{\mathbf{A}}_{11} + \mathbf{P}\bar{\mathbf{A}}_{21})\mathbf{P} = 0, \tag{4.38}$$

iteratively for sufficiently small value of $\mu$ (Naidu 1988). Consequently, the system (4.37) is modeled into lower triangular form, where $\mathbf{z}_s$ is completely decoupled from $\mathbf{z}_b$ and it is given by

$$\dot{\mathbf{g}}_s = \bar{\mathbf{A}}_s \mathbf{z}_s + \bar{\mathbf{B}}_s \mathbf{u}. \tag{4.39}$$

Recall that, $\mathbf{z}_s$ combines $(n_1 + p)$ 'slow' and $n_2$ 'fast' states of the system (4.32). Hence, it can also be written in two-time-scale form, by partitioning $\bar{\mathbf{A}}_s$ and $\bar{\mathbf{B}}_s$ suitably, as

$$\begin{bmatrix} \dot{\mathbf{z}}_{s1} \\ \dot{\mathbf{z}}_{s2} \end{bmatrix} = \begin{bmatrix} \bar{\mathbf{A}}_{s11} & \bar{\mathbf{A}}_{s12} \\ \frac{\bar{\mathbf{A}}_{s21}}{\varepsilon} & \frac{\bar{\mathbf{A}}_{s22}}{\varepsilon} \end{bmatrix} \begin{bmatrix} \mathbf{z}_{s1} \\ \mathbf{z}_{s2} \end{bmatrix} + \begin{bmatrix} \bar{\mathbf{B}}_{s1} \\ \frac{\bar{\mathbf{B}}_{s2}}{\varepsilon} \end{bmatrix} \mathbf{u}, \tag{4.40}$$

where $\mathbf{z}_s = \left[\mathbf{z}_{s1}^T \ \mathbf{z}_{s2}^T\right]^T$ with $\mathbf{z}_{s1} \in \Re^{n_1+p}$ and $\mathbf{z}_{s2} \in \Re^{n_2}$. At this moment, a linear transformation

$$\mathbf{z}_{ss} = \mathbf{z}_{s1} + \varepsilon \mathbf{P}_s \mathbf{z}_{s2}, \tag{4.41}$$

is applied to (4.40). Here, $\mathbf{P}_s$ is determined by solving

$$\bar{\mathbf{A}}_{s12} + \mathbf{P}_s \bar{\mathbf{A}}_{s22} - \varepsilon(\bar{\mathbf{A}}_{s11} + \mathbf{P}_s \bar{\mathbf{A}}_{s21})\mathbf{P}_s = 0. \tag{4.42}$$

Solution of $\mathbf{P}_s$ is obtained under Assumption 4.3 that $\mathbf{A}_{22}$, i.e., $\bar{\mathbf{A}}_{s22}$ is invertible for a small value of $\varepsilon$. Accordingly, a lower triangular form of (4.40) is obtained as

$$\begin{bmatrix} \dot{\mathbf{z}}_{ss} \\ \dot{\mathbf{z}}_{s2} \end{bmatrix} = \begin{bmatrix} \bar{\mathbf{A}}_{ss} & \mathbf{0} \\ \frac{\bar{\mathbf{A}}_{s21}}{\varepsilon} & \frac{\bar{\mathbf{A}}_{sf}}{\varepsilon} \end{bmatrix} \begin{bmatrix} \mathbf{z}_{s1} \\ \mathbf{z}_{s2} \end{bmatrix} + \begin{bmatrix} \bar{\mathbf{B}}_{ss} \\ \frac{\bar{\mathbf{B}}_{s2}}{\varepsilon} \end{bmatrix} \mathbf{u}, \tag{4.43}$$

where $\bar{\mathbf{A}}_{ss} = \bar{\mathbf{A}}_{s11} + \mathbf{P}_s \bar{\mathbf{A}}_{s21}$, $\bar{\mathbf{A}}_{sf} = \bar{\mathbf{A}}_{s22} - \varepsilon \bar{\mathbf{A}}_{s21} \mathbf{P}_s$ and $\bar{\mathbf{B}}_{ss} = \bar{\mathbf{B}}_{s1} + \mathbf{P_s} \bar{\mathbf{B}}_{s2}$. Thus, using block-triangular forms (4.37) and (4.43) of two-time-scale representations, the original three-time-scale system (4.32) is converted into block-triangular form

$$\begin{bmatrix} \dot{\mathbf{z}}_{ss} \\ \dot{\mathbf{z}}_{s2} \\ \dot{\mathbf{z}}_{3} \end{bmatrix} = \begin{bmatrix} \bar{\mathbf{A}}_{ss} & \mathbf{0} & \mathbf{0} \\ \frac{\bar{\mathbf{A}}_{s21}}{\varepsilon} & \frac{\bar{\mathbf{A}}_{sf}}{\varepsilon} & \mathbf{0} \\ \frac{\bar{\mathbf{A}}_{31}}{\mu} & \frac{\bar{\mathbf{A}}_{32}}{\mu} & \frac{\bar{\mathbf{A}}_{f}}{\mu} \end{bmatrix} \begin{bmatrix} \mathbf{z}_{ss} \\ \mathbf{z}_{s2} \\ \mathbf{z}_{3} \end{bmatrix} + \begin{bmatrix} \bar{\mathbf{B}}_{ss} \\ \frac{\bar{\mathbf{B}}_{s2}}{\varepsilon} \\ \frac{\bar{\mathbf{B}}_{2}}{\mu} \end{bmatrix} \mathbf{u}, \tag{4.44}$$

where $\bar{\mathbf{A}}_{21}$ in (4.37) is appropriately partitioned as $[\bar{\mathbf{A}}_{31} \quad \bar{\mathbf{A}}_{32}]$.

Now, one can proceed for a three-stage feedback control design starting with the 'slow' subsystem $\mathbf{z}_{ss}$. To begin with, the control is designed for the system (4.40) and then for the system (4.35). In the first stage, applying feedback control

$$\mathbf{u} = -\mathbf{K}_{ss} \mathbf{z}_{ss} + \mathbf{u}_{sf}, \tag{4.45}$$

to the system (4.43), leads to

$$\begin{bmatrix} \dot{\mathbf{z}}_{ss} \\ \dot{\mathbf{z}}_{s2} \end{bmatrix} = \begin{bmatrix} \bar{\mathbf{A}}_{ss} - \bar{\mathbf{B}}_{ss} \mathbf{K}_{ss} & \mathbf{0} \\ \frac{\bar{\mathbf{A}}_{s21} - \bar{\mathbf{B}}_{s2} \mathbf{K}_{ss}}{\varepsilon} & \frac{\bar{\mathbf{A}}_{sf}}{\varepsilon} \end{bmatrix} \begin{bmatrix} \mathbf{z}_{ss} \\ \mathbf{z}_{s2} \end{bmatrix} + \begin{bmatrix} \bar{\mathbf{B}}_{ss} \\ \frac{\bar{\mathbf{B}}_{s2}}{\varepsilon} \end{bmatrix} \mathbf{u}_{sf}, \tag{4.46}$$

where $\mathbf{K}_{ss}$ is selected such that $\psi(\bar{\mathbf{A}}_{ss} - \bar{\mathbf{B}}_{ss} \mathbf{K}_{ss})$ are 'slow' and in left half of $s$-plane. Now, using the change of the variables

$$\mathbf{z}_{sf} = -\mathbf{M}_s \mathbf{z}_{ss} + \mathbf{z}_{s2}, \tag{4.47}$$

the system (4.46) is modified to block-diagonal form

$$\begin{bmatrix} \dot{\mathbf{z}}_{ss} \\ \dot{\mathbf{z}}_{sf} \end{bmatrix} = \begin{bmatrix} \bar{\mathbf{A}}_{ss} - \bar{\mathbf{B}}_{ss} \mathbf{K}_{ss} & \mathbf{0} \\ \mathbf{0} & \frac{\bar{\mathbf{A}}_{sf}}{\varepsilon} \end{bmatrix} \begin{bmatrix} \mathbf{z}_{ss} \\ \mathbf{z}_{sf} \end{bmatrix} + \begin{bmatrix} \bar{\mathbf{B}}_{ss} \\ \frac{\bar{\mathbf{B}}_{sf}}{\varepsilon} \end{bmatrix} \mathbf{u}_{sf}, \tag{4.48}$$

where $\bar{\mathbf{B}}_{sf} = \bar{\mathbf{B}}_{s2} - \varepsilon \mathbf{M}_s \bar{\mathbf{B}}_{ss}$ and matrix $\mathbf{M}_s$ satisfies

$$\bar{\mathbf{A}}_{s21} - \bar{\mathbf{B}}_{s2} \mathbf{K}_{ss} - \varepsilon \mathbf{M}_s (\bar{\mathbf{A}}_{ss} - \bar{\mathbf{B}}_{ss} \mathbf{K}_{ss}) + \bar{\mathbf{A}}_{sf} \mathbf{M}_s = 0. \tag{4.49}$$

Solution of (4.49) exists if Assumption 4.4 is satisfied.

**Assumption 4.4** $\psi(\bar{\mathbf{A}}_{ss} - \bar{\mathbf{B}}_{ss}\mathbf{K}_{ss}) \neq \psi\left(\frac{\bar{\mathbf{A}}_{sf}}{\varepsilon}\right)$.

It is satisfied because $(\bar{\mathbf{A}}_{ss} - \bar{\mathbf{B}}_{ss}\mathbf{K}_{ss})$ is user designed and has desired 'slow' eigenvalues, while $\frac{\bar{\mathbf{A}}_{sf}}{\varepsilon}$ has original 'fast' subsystem eigenvalues. In (4.48), subsystem $\mathbf{z}_{sf}$ is made totally independent of $\mathbf{z}_{ss}$. Thus, in the second stage

$$\mathbf{u}_{sf} = -\mathbf{K}_{sf}\mathbf{z}_{sf}, \tag{4.50}$$

is applied to the system (4.48) as

$$\begin{bmatrix} \dot{\mathbf{z}}_{ss} \\ \dot{\mathbf{z}}_{sf} \end{bmatrix} = \begin{bmatrix} \bar{\mathbf{A}}_{ss} - \bar{\mathbf{B}}_{ss}\mathbf{K}_{ss} & -\bar{\mathbf{B}}_{ss}\mathbf{K}_{sf} \\ \mathbf{0} & \frac{\bar{\mathbf{A}}_{sf} - \bar{\mathbf{B}}_{sf}\mathbf{K}_{sf}}{\varepsilon} \end{bmatrix} \begin{bmatrix} \mathbf{z}_{ss} \\ \mathbf{z}_{sf} \end{bmatrix}, \tag{4.51}$$

so that $\psi\left(\frac{\bar{\mathbf{A}}_{sf} - \bar{\mathbf{B}}_{sf}\mathbf{K}_{sf}}{\varepsilon}\right)$ are made asymptotically stable. Thus, the overall input for the system (4.40) is computed using transformations (4.41) and (4.47) and controls (4.45) and (4.50) as

$$\mathbf{u} = -\mathbf{K}_s \left[\mathbf{z}_{s1}^T \ \mathbf{z}_{s2}^T\right]^T = -\mathbf{K}_s\mathbf{z}_s, \tag{4.52}$$

where

$$\mathbf{K}_s = \left[\mathbf{K}_{ss} - \mathbf{K}_{sf}\mathbf{M}_s \ \ \varepsilon(\mathbf{K}_{ss} - \mathbf{K}_{sf}\mathbf{M}_s)\mathbf{P}_s + \mathbf{K}_{sf}\right]. \tag{4.53}$$

Actually, (4.52) is the input for (4.39), which is the decoupled part of (4.37). So, the overall input (4.52) has the form

$$\mathbf{u} = -\mathbf{K}_s\mathbf{z}_s + \mathbf{u}_f, \tag{4.54}$$

and is passed on to the system (4.37). This results in

$$\begin{bmatrix} \dot{\mathbf{z}}_s \\ \dot{\mathbf{z}}_b \end{bmatrix} = \begin{bmatrix} \bar{\mathbf{A}}_s - \bar{\mathbf{B}}_s\mathbf{K}_s & \mathbf{0} \\ \frac{\bar{\mathbf{A}}_{21} - \bar{\mathbf{B}}_2\mathbf{K}_s}{\mu} & \frac{\bar{\mathbf{A}}_f}{\mu} \end{bmatrix} \begin{bmatrix} \mathbf{z}_s \\ \mathbf{z}_b \end{bmatrix} + \begin{bmatrix} \bar{\mathbf{B}}_s \\ \frac{\bar{\mathbf{B}}_2}{\mu} \end{bmatrix} \mathbf{u}_f. \tag{4.55}$$

Once again introducing a change of coordinates as

$$\mathbf{z}_f = -\mathbf{M}\mathbf{z}_s + \mathbf{z}_b, \tag{4.56}$$

for the system (4.55), subsystem $\mathbf{z}_f$ is isolated from $\mathbf{z}_s$ as

$$\begin{bmatrix} \dot{\mathbf{z}}_s \\ \dot{\mathbf{z}}_f \end{bmatrix} = \begin{bmatrix} \bar{\mathbf{A}}_s - \bar{\mathbf{B}}_s\mathbf{K}_s & \mathbf{0} \\ \mathbf{0} & \frac{\bar{\mathbf{A}}_f}{\mu} \end{bmatrix} \begin{bmatrix} \mathbf{z}_s \\ \mathbf{z}_f \end{bmatrix} + \begin{bmatrix} \bar{\mathbf{B}}_s \\ \frac{\bar{\mathbf{B}}_f}{\mu} \end{bmatrix} \mathbf{u}_f, \tag{4.57}$$

where $\bar{\mathbf{B}}_f = \bar{\mathbf{B}}_2 - \mu\mathbf{M}\bar{\mathbf{B}}_s$ and $\mathbf{M}$ is evaluated by setting

$$\bar{\mathbf{A}}_{21} - \bar{\mathbf{B}}_2\mathbf{K}_s - \mu\mathbf{M}(\bar{\mathbf{A}}_s - \bar{\mathbf{B}}_s\mathbf{K}_s) + \bar{\mathbf{A}}_f\mathbf{M} = 0. \tag{4.58}$$

Assumption 4.5 is needed to solve (4.58).

**Assumption 4.5** $\psi(\bar{\mathbf{A}}_s - \bar{\mathbf{B}}_s\mathbf{K}_s) \neq \psi\left(\frac{\bar{\mathbf{A}}_f}{\mu}\right)$.

And, it can be convinced by the fact that $(\bar{\mathbf{A}}_s - \bar{\mathbf{B}}_s\mathbf{K}_s)$ is designed with desired eigenvalues laying in 'slow' and 'fast' dynamic modes and $\frac{\bar{\mathbf{A}}_f}{\mu}$ has 'very fast' eigenvalues of full-order system. Finally, in the third stage

$$\mathbf{u}_f = -\mathbf{K}_f\mathbf{z}_f, \tag{4.59}$$

is applied to the system (4.57) as

$$\begin{bmatrix} \dot{\mathbf{z}}_s \\ \dot{\mathbf{z}}_f \end{bmatrix} = \begin{bmatrix} \bar{\mathbf{A}}_s - \bar{\mathbf{B}}_s\mathbf{K}_s & -\bar{\mathbf{B}}_s\mathbf{K}_f \\ \mathbf{0} & \frac{\bar{\mathbf{A}}_f - \bar{\mathbf{B}}_f\mathbf{K}_f}{\mu} \end{bmatrix} \begin{bmatrix} \mathbf{z}_s \\ \mathbf{z}_f \end{bmatrix}, \tag{4.60}$$

so as to place $\psi\left(\frac{\bar{\mathbf{A}}_f - \bar{\mathbf{B}}_f\mathbf{K}_f}{\mu}\right)$ at desired locations. Therefore, the overall input for the system (4.35), consequently for system (4.32), can be obtained from (4.52), (4.54), (4.56) and (4.59) as

$$\mathbf{u} = -\hat{\mathbf{K}}\left[\mathbf{z}_a^T \ \mathbf{z}_b^T\right]^T = -\hat{\mathbf{K}}\mathbf{z}_{aug}, \tag{4.61}$$

where

$$\hat{\mathbf{K}} = \left[\mathbf{K}_s - \mathbf{K}_f\mathbf{M} \ \ \mu(\mathbf{K}_s - \mathbf{K}_f\mathbf{M})\mathbf{P} + \mathbf{K}_f\right]. \tag{4.62}$$

This is the composite feedback gain, derived from subsystem feedback gains designed independently, so that, the original system (4.16) is asymptotically stable, i.e., $\psi(\hat{\mathbf{A}} - \hat{\mathbf{B}}\hat{\mathbf{K}}) < 0$.

**Remark 4.1** Interestingly, only two nonsymmetric Riccati equations (4.38) and (4.42) respectively for $\mathbf{P}$ and $\mathbf{P}_s$ need to be evaluated to obtain the lower triangular form (4.44) of (4.32), compared to three equations in (Radisavljevic-Gajic et al. 2017). Further, to execute a three-stage design, the solutions of just two Sylvester equations (4.49) and (4.58) respectively for $\mathbf{M}_s$ and $\mathbf{M}$ are needed, as opposed to three equations in Radisavljevic-Gajic et al. (2017). The fixed point iteration method presented in Radisavljevic-Gajic et al. (2017) can be utilized to determine the solutions of these algebraic equations. (ii) Although (Radisavljevic-Gajic et al. 2017) contains six equations that must be solved, only four algebraic equations-four of which are of higher order-need to be solved. Therefore, for both approaches, the floating point operations (FLOPs) needed to obtain the answer are calculated and compared. The

FLOPs are expressed as a function of the matrix dimension. If $n_1$, $n_2$ and $n_3$ are the orders of the slow, fast and very fast subsystems, inequality

$$2n^3 + 6h^3 + \frac{2}{3}n_3^3 + n^2 + n_3^2 > 2h^2(1 + 4n) + 4n_1n_2n_3, \tag{4.63}$$

is derived, in which $n = n_1 + n_2 + n_3$ and $h = n_1 + n_2$. It is noticed that the online computational requirements will be reduced by the proposed method if the inequality (4.63) is satisfied. (iii) Besides, the offline computational requirements are significantly reduced as calculations are done with the (2×2) matrix representation instead of (3×3) matrix representation, without the loss of accuracy and the freedom to design different local feedback controllers for individual subsystems.

**Remark 4.2** If $\psi\left(\frac{\bar{\mathbf{A}}_f}{\mu}\right) < 0$, only 'slow' and 'fast' subsystems are needed to be stabilized by corresponding controls (4.45) and (4.50). Accordingly, $\mathbf{K}_f$ can be set to zero in (4.62), resulting in $\bar{\mathbf{K}} = \left[\mathbf{K}_s \;\; \mu\mathbf{K}_s\mathbf{P}\right]$, where $\mathbf{K}_s$ is given by (4.53).

**Remark 4.3** In the presence of $\psi\left(\frac{\bar{\mathbf{A}}_{sf}}{\varepsilon}\right) < 0$ and $\psi\left(\frac{\bar{\mathbf{A}}_f}{\mu}\right) < 0$, 'slow' subsystem alone demands stabilization as per control (4.45). As a deduction, both $\mathbf{K}_{sf}$ and $\mathbf{K}_f$ can be replaced with zeros in (4.53) and (4.62) respectively, ending with gain as $\tilde{\mathbf{K}} = \left[\tilde{\mathbf{K}}_s \;\; \mu\tilde{\mathbf{K}}_s\mathbf{P}\right]$, where $\tilde{\mathbf{K}}_s = [\mathbf{K}_{ss} \;\; \varepsilon\mathbf{K}_{ss}\mathbf{P}_s]$.

When formulating a controller according to Remark 4.2, one has to deal with three algebraic equations, i.e., for $\mathbf{P}$, $\mathbf{P}_s$, and $\mathbf{M}_s$. However, when designing a controller utilizing Remark 4.3, one needs only deal with two equations, i.e., for $\mathbf{P}$ and $\mathbf{P}_s$. This reduces computations even more.

#### 4.3.2.2 Design of Observer Gain (L)

Designing an observer gain for (4.13)–(4.14) is analogous to solving an eigenvalue assignment problem for the dual system

$$\dot{\mathbf{x}} = \mathbf{A}^T\mathbf{x} + \mathbf{C}^T\nu, \tag{4.64}$$

$$\xi = \mathbf{B}^T\mathbf{x}, \tag{4.65}$$

with control $\nu = \mathbf{L}^T\mathbf{x}$, such that $\psi(\mathbf{A} - \mathbf{LC}) < 0$, provided Assumption 4.1 is fulfilled. Note that, dual system (4.64)–(4.65) also displays three-time-scale property, because $\psi(\mathbf{A}) = \psi(\mathbf{A}^T)$. And so, for the suggested three-stage observer design consider an analogous system to (4.30)–(4.31) (Yoo and Gajic 2018) as

$$\begin{bmatrix} \dot{\mathbf{x}}_1 \\ \dot{\mathbf{x}}_2 \\ \dot{\mathbf{x}}_3 \end{bmatrix} = \begin{bmatrix} \mathbf{A}_{o11}^T & \frac{\mathbf{A}_{o21}^T}{\varepsilon} & \frac{\mathbf{A}_{o31}^T}{\mu} \\ \mathbf{A}_{o12}^T & \frac{\mathbf{A}_{o22}^T}{\varepsilon} & \frac{\mathbf{A}_{o32}^T}{\mu} \\ \mathbf{A}_{o13}^T & \frac{\mathbf{A}_{o23}^T}{\varepsilon} & \frac{\mathbf{A}_{o33}^T}{\mu} \end{bmatrix} \begin{bmatrix} \mathbf{x}_1 \\ \mathbf{x}_2 \\ \mathbf{x}_3 \end{bmatrix} + \begin{bmatrix} \mathbf{C}_{o1}^T \\ \mathbf{C}_{o2}^T \\ \mathbf{C}_{o3}^T \end{bmatrix} \nu. \tag{4.66}$$

Introducing

$$\begin{bmatrix} \mathbf{q}_1 \\ \mathbf{q}_2 \\ \mathbf{q}_3 \end{bmatrix} = \begin{bmatrix} \mathbf{I}_{n1} & \mathbf{0} & \mathbf{0} \\ \mathbf{0} & \frac{\mathbf{I}_{n2}}{\varepsilon} & \mathbf{0} \\ \mathbf{0} & \mathbf{0} & \frac{\mathbf{I}_{n3}}{\mu} \end{bmatrix} \begin{bmatrix} \mathbf{x}_1 \\ \mathbf{x}_2 \\ \mathbf{x}_3 \end{bmatrix} = \mathbf{T}^T \begin{bmatrix} \mathbf{x}_1 \\ \mathbf{x}_2 \\ \mathbf{x}_3 \end{bmatrix}, \tag{4.67}$$

standard singularly perturbed three-time-scale form of (4.66) is obtained as

$$\begin{bmatrix} \dot{\mathbf{q}}_1 \\ \dot{\mathbf{q}}_2 \\ \dot{\mathbf{q}}_3 \end{bmatrix} = \begin{bmatrix} \mathbf{A}_{o11}^T & \mathbf{A}_{o21}^T & \mathbf{A}_{o31}^T \\ \frac{\mathbf{A}_{o12}^T}{\varepsilon} & \frac{\mathbf{A}_{o22}^T}{\varepsilon} & \frac{\mathbf{A}_{o32}^T}{\varepsilon} \\ \frac{\mathbf{A}_{o13}^T}{\mu} & \frac{\mathbf{A}_{o23}^T}{\mu} & \frac{\mathbf{A}_{o33}^T}{\mu} \end{bmatrix} \begin{bmatrix} \mathbf{q}_1 \\ \mathbf{q}_2 \\ \mathbf{q}_3 \end{bmatrix} + \begin{bmatrix} \mathbf{C}_{o1}^T \\ \frac{\mathbf{C}_{o2}^T}{\varepsilon} \\ \frac{\mathbf{C}_{o3}^T}{\mu} \end{bmatrix} \nu, \tag{4.68}$$

where $\mathbf{q}_1 \in \Re^{n_1}$, $\mathbf{q}_2 \in \Re^{n_2}$ and $\mathbf{q}_3 \in \Re^{n_3}$. The construction of the lower triangular form of (4.68) and the exercise of the three stages of design for an observer gain can be done identically to that of the feedback gain, as discussed in Sect. 4.3.2.1 from (4.34) to (4.62). This will yield $\mathbf{L}_{ss}$, $\mathbf{L}_{sf}$ and $\mathbf{L}_f$ after each stage, which correspond to $\mathbf{K}_{ss}$, $\mathbf{K}_{sf}$ and $\mathbf{K}_f$, respectively. The observer gain is intended for the original system (4.30)–(4.31), whereas the feedback gain is intended for the augmented system (4.32)–(4.33). This means that the differences lie in the dimensions of the state transformations. In addition, system (4.67) should be used to represent the final composite observer gain in the original states. Designing subsystem observer gains requires careful consideration of Assumptions 4.3–4.5.

## 4.4 Application to CFPP

This section first discusses the control design considerations, and then it applies those aspects to a coal-fired 300 MW power station under representative transient conditions.

### *4.4.1 Controller Design*

Section 4.2 confirms the 300 MW CFPP exhibits a three-time-scale structure. It has six states and two outputs. The augmented system (4.16) becomes of order eight when an integral control is applied. Thus, Theorem 4.1 states that the order of the 'slow' mode increases from 2 to 4 with eigenvalues of $(0, 0, -0.0026, -0.0071)$ respectively. The positions of the eigenvalues of the 'fast' and 'very fast' modes do not appear to have changed. Additionally, ratios $\varepsilon$ and $\mu$ do not change. Because of stable 'fast' and 'very fast' states, the next step is to acquire total state feedback gain using only 'slow' subsystem states. It is noted that the CFPP augmented model exists in an implicit version. Hence, it is transformed into explicitly singularly perturbed form by replacing some rows of the system matrix such that inverses of $\mathbf{A}_{22}$ and $\mathbf{A}_{33}$ exist (Naidu 1988). After that, with suggested divisions and mergers of states, the system (4.38) is solved to get $\mathbf{P}$ which helps to find $\bar{\mathbf{A}}_f$ and $\bar{\mathbf{A}}_s$. Here, one can affirm that the eigenvalues of $\bar{\mathbf{A}}_f$ are the same as that of 'very fast' mode, whereas eigenvalues of $\bar{\mathbf{A}}_s$ are the union of 'slow' and 'fast' modes. Now solving (4.42) provides $\mathbf{P}_s$. Resultantly, one can gain $\bar{\mathbf{A}}_{ss}$ and $\bar{\mathbf{A}}_{sf}$. Again, it is checked that the eigenvalues of $\bar{\mathbf{A}}_{ss}$ and $\bar{\mathbf{A}}_{sf}$ are equivalent to the eigenvalues of 'slow' and 'fast' modes respectively. Considering no requirement of control for 'fast' and 'very fast' modes, the desired eigenvalues of the 'slow' subsystem are selected as $(-0.97, -0.86, -0.75, -0.64) \times 10^{-3}$, and these are accurately placed by determining feedback gain $\mathbf{K}_{ss}$. Substituting $\mathbf{K}_{sf} = 0$, feedback gain at the end of 2nd stage is obtained and again making $\mathbf{K}_f = 0$, one can have $\tilde{\mathbf{K}}$, after 3rd stage. The closed-loop eigenvalues, i.e., $\psi(\hat{\mathbf{A}} - \hat{\mathbf{B}}\tilde{\mathbf{K}})$ are found to be precisely placed at

$$\psi^{desired} = \big[-0.00064 \; -0.00075 \; -0.00086 \; -0.00097 \\ -0.0454 \; -0.1306 \pm j0.0361 \; -0.2497\big].$$

The feedback gain has a maximum value of $-19.0392$, which appears to be practically implementable. It is also computed to set the eigenvalues at $\psi^{desired}$ using the direct design of feedback gain. In order to put observer poles at $(-3.7455, -1.9590 \pm j0.5415, -0.0683, -0.0146, -0.0129)$, observer gains are calculated for (4.13) using both three-stage design and direct design. This is done to implement a controller. It is observed that the feedback and observer gain values acquired via the process of Radisavljevic-Gajic et al. (2017) are identical. Conversely, it is noticed that there are more FLOPs needed because the inequality (4.63) is satisfied. The suggested technique requires 1092 FLOPs for determining feedback and observer gains, but Radisavljevic-Gajic et al. (2017) states that FLOPs of 1259 and 711 are needed, respectively.

### *4.4.2 Transient Simulations*

Feedback gains obtained by the suggested three-stage design and direct design are utilized to generate simulation results. These results are then compared with a direct energy balance-based PI (DEB-PI) controller (Sun et al. 2016). It is made sure the plant (4.1)–(4.8) is in a steady state before starting any transient. First, to assess their convergence, estimation errors with both observer gains (i.e., three-stage and direct design) are compared. It has been noted that the three-stage design convergence performs better than the direct design. As a result, the observer gain from the three-stage architecture is applied in every scenario.

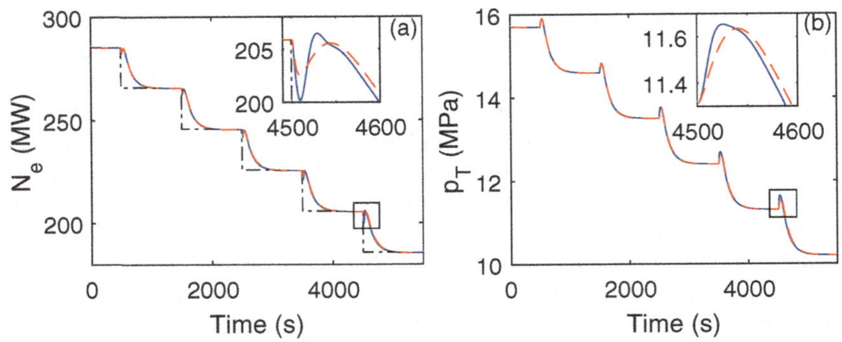


**Fig. 4.1** Effect of change in operating conditions from 285.9 to 185.9 MW on **a** Electrical power and **b** Throttle pressure (Black in dashed-dotted: Setpoint, blue in solid: Three-stage design, red in dashed: DEB-PI)

**Table 4.1** Comparison of three-stage design and DEB-PI

| Operating conditions (MW) | | 285.9–265.9 | 265.9–245.9 | 245.9–225.9 | 225.9–205.9 | 205.9–185.9 |
|---|---|---|---|---|---|---|
| IAE | Three-Stage | 118.27 | 118.48 | 118.68 | 118.90 | 119.13 |
| | DEB-PI | 118.85 | 120.24 | 121.79 | 123.51 | 125.48 |
| ISE | Three-Stage | 1487.9 | 1491.9 | 1496.0 | 1501.0 | 1508.3 |
| | DEB-PI | 1490.7 | 1511.5 | 1536.2 | 1566.6 | 1604.8 |

#### 4.4.2.1 Tracking Performance

In this case, as Fig. 4.1a illustrates, the operating condition is gradually reduced from 285.9 MW by 20 MW every 1000 s, beginning at 500 s. Figure 4.1b displays corresponding changes in throttle pressure. It is observed that the response caused by directly designed feedback control begins to oscillate with increasing amplitude as the operating point decreases and becomes unstable after 225.9 MW. Therefore, Fig. 4.1 does not display it. The overall response, however, was found to be steady with a three-stage design with DEB-PI, with a tracking rate of output power of 1.6% of the full load per minute (Fig. 4.1a). Furthermore, the throttle pressure's reverse response is examined to ensure that it is within $\pm$ 0.4 MPa (refer to the inset of Fig. 4.1b). Since the responses from the DEB-PI and the suggested control are similar, the error indices for power, as listed in Table 4.1, are computed to compare the two sets of findings. The improvement in the suggested three-stage architecture over DEB-PI is seen in Table 4.1.

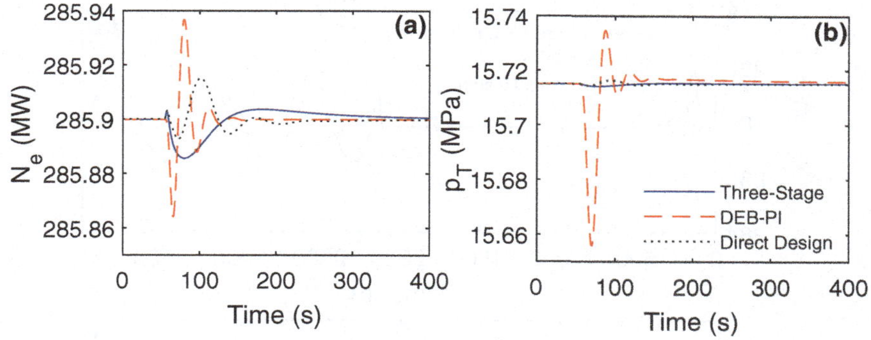


**Fig. 4.2** Effect of 5% variation in boiler demand on **a** Electrical power and **b** Throttle pressure

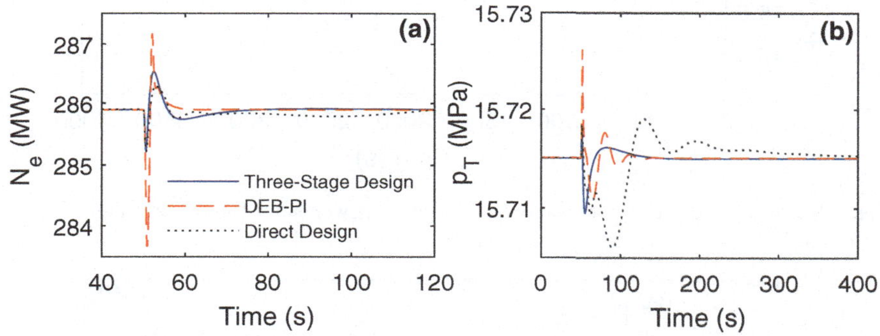


**Fig. 4.3** Effect of 5% variation in throttle opening position on **a** Electrical power and **b** Throttle pressure

#### 4.4.2.2 Regulating Performance

First, after 50 s, the boiler demand is reduced by 5% of its nominal value and then increased again. The impact of this on the throttle pressure and output power is monitored in Fig. 4.2. Second, after 50 s, a throttle opening position is dropped by 5% and then recovered. The impact of this on pressure and power is shown in Fig. 4.3. In all scenarios, every controller effectively controls pressure and power. When compared to the other two controllers, the suggested three-stage architecture is found to have a better overall response.

#### 4.4.2.3 Disturbance Rejection and Robustness

This study examines the impact of variations in coal quality, starting with a 20% rapid rise at 100 s and continuing with a constant change lasting 628 s after 2100 s. All of the controllers control the power and pressure during the sudden change, as shown in

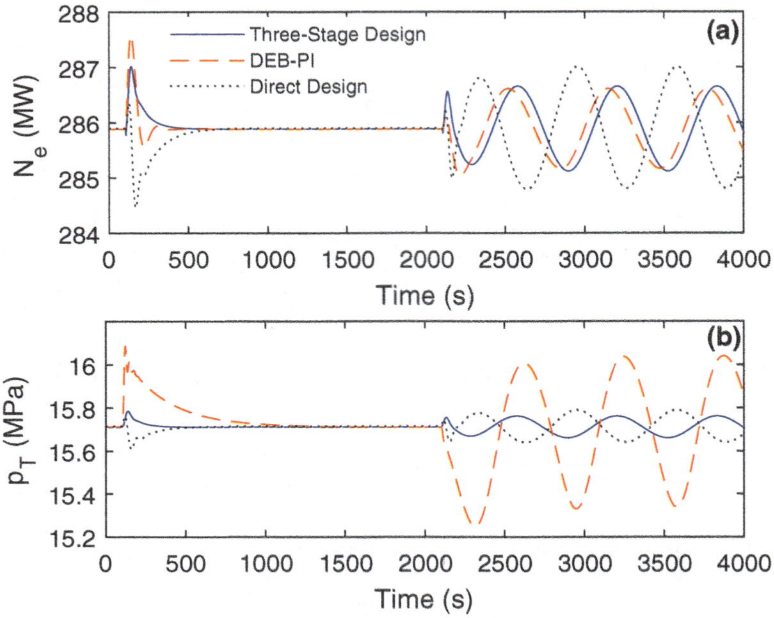


**Fig. 4.4** Effect of change in coal quality on **a** Electrical power and **b** Throttle pressure

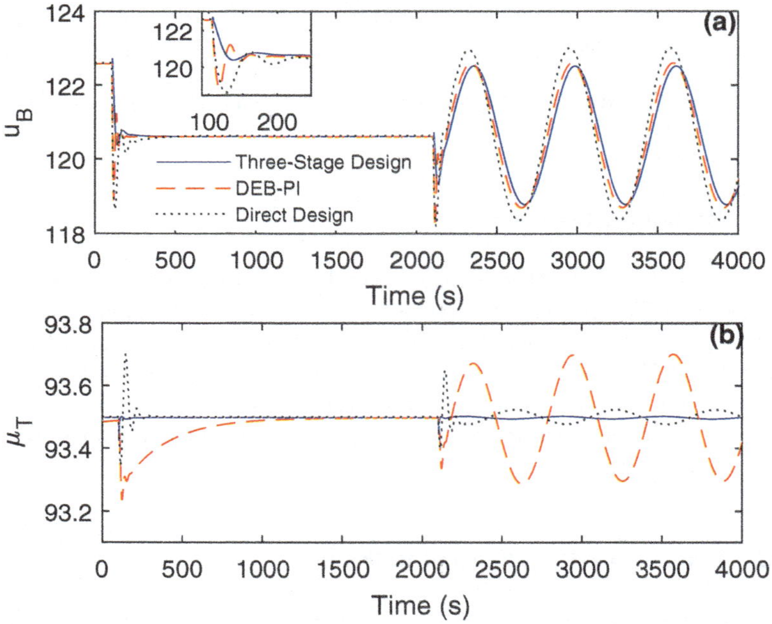


**Fig. 4.5** Effect of change in coal quality on control inputs, i.e., **a** Boiler demand (t/h) and **b** Throttle opening position (%)

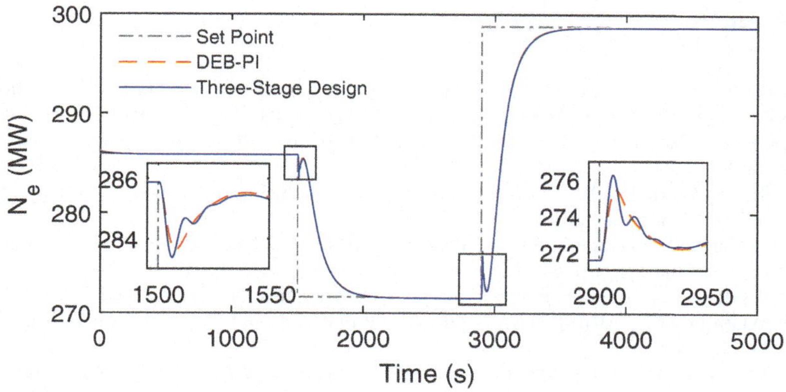


**Fig. 4.6** Effect of parameter uncertainty on electrical power

Fig. 4.4. Nonetheless, it is found that the three-stage design yields a superior response than both direct design and DEB-PI. When it comes to periodic variations, pressure exhibits oscillations of a smaller magnitude while power exhibits oscillations of a lower magnitude as those produced with DEB-PI. Corresponding variations in control inputs, displayed in Fig. 4.5, also indicate performance improvement by three-stage design over the direct design and DEB-PI. In the end, the set-point changes with the change in the dynamic constants given in Sun et al. (2016) are observed to evaluate the resilience of controllers for parameter uncertainty. In this case, the three-stage design and DEB-PI responded as shown in Fig. 4.6, while the control resulting from the direct design produced an unstable response. The insets of Fig. 4.6 display the instants of set-point initiations.

## 4.5 Conclusion

This chapter proposes an observer-based output feedback control employing an integral action for a three-time-scale system, where the two-time-scale technique is used to construct both the feedback and the observer gains in three phases. The three-stage design approach eliminates the higher order multi-time-scale system's ill-conditioning concerns and gives significant online and offline computation reductions. The nonlinear simulations of a large-scale 300 MW CFPP demonstrate the benefits and efficacy of the proposed controller. Comparing the suggested controller to the direct design approach and the direct energy balance-based PI controller, simulations show that it achieves significant tracking and regulating control in the presence of input and parameter perturbations and meets the demanding control requirements of the real-world power plant.

## References

Alamoodi N, Daoutidis P (2016) Nonlinear decoupling control with deadtime compensation for multirange operation of steam power plants. IEEE Trans Control Syst Technol 24:341–348
Alamoodi N, Daoutidis P (2017) Nonlinear control of coal fired steam power plants. Control Eng Pract 60:63–75
Annual Energy Outlook (2017) with Projections to 2050. Energy Information Administration, Washington, DC, USA, U.S, p 2017
Chandrasekharan S, Panda RC, Swaminathan BN (2014) Modeling, identification, and control of coal-fired thermal power plants. Rev Chem Eng 30:217–232
Chen CT (1999) Linear system theory and design. Oxford University Press, New York
Cori R, Maffezzoni C (2016) Practical-optimal control of a drum boiler power plant. Automatica 20:163–173
Gamal M, Banna ME, Farag MM (2016) Hardware implementation of an LQR controller of a drum-type boiler turbine on FPGA. In: Proceedings of the IEEE international conference microelectronics, pp 137–140
Key World Energy Statistics 2017. International Energy Agency, Paris, France (2017)
Li S, Liu H, Cai W-J, Soh Y-C, Xie LH (2005) A new coordinated control strategy for boiler-turbine system of coal-fired power plant. IEEE Trans Control Syst Technol 13:943–954
Liu X, Guan P, Chan CW (2010) Nonlinear multivariable power plant coordinate control by constrained predictive scheme. IEEE Trans Control Syst Technol 18:1116–1125
Moradi H, Vossoughi G (2016) Multivariable optimal control of an industrial nonlinear boiler-turbine unit. Meccanica 51:859–875
Munje RK, Lin S, Zhang G, Zhang W (2020) Observer-based output feedback integral control for coal-fired power plant: a three- time-scale perspective. IEEE Trans Control Syst Technol 28:601–608
Naidu DS, Ravinder R (1985) On three time scale analysis. In: Proceedings of the conference on decision and control, pp 81–85
Naidu DS (1988) Singular perturbation methodology in control systems. Peter Peregrinus Ltd, London U.K
Naidu DS, Rao AK (1985) Application of the singular perturbation method to a steam power system. Electr Power Syst Res 8:219–226
Patre BM, Bandyopadhyay B, Werner H (1999) Periodic output feedback control for singularly perturbed discrete model of a steam power system. Proc IEE Control Theory Appl 146:247–252
Peng H, Ozaki T, Haggn-Ozaki V, Toyoda Y (2002) A nonlinear exponential ARX model-based multivariable generalized predictive control strategy for thermal power plants. IEEE Trans Control Syst Technol 10:256–262
Phillips R (1980) A two-stage design of linear feedback controls. IEEE Trans Auto Control 25:1220–1223
Radisavljevic-Gajic V, Rose P (2014) A new two-stage design of feedback controllers for a hydrogen gas reformer. Int J Hydrog Energy 39:11738–11748
Radisavljevic-Gajic V, Milanovic M, Clayton G (2017) Three-stage feedback controller design with application to three-time-scale linear control system. ASME J Dyn Syst, Meas, Control 139:1–10
Singh NP, Singh YP, Ahson SI (1986) An iterative approach to reduced- order modeling of synchronous machines using singular perturbation. Proc IEEE 74:892–893
Sun L, Li D, Lee KY, Xue Y (2016) Control-oriented modeling and analysis of direct energy balance in coal-fired boiler-turbine unit. Control Eng Pract 55:38–55
Sun L, Hua Q, Li D, Pan L, Xue Y, Lee KY (2017) Direct energy balance based active disturbance rejection control for coal-fired power plant. ISA Trans 70:486–493
Wei L, Fang F (2017) H$\infty$-LQR-based coordinated control for large coal-fired boiler turbine generation units. IEEE Trans Ind Electron 64:5212–5221
Wu X, Shen J, Li Y, Lee KY (2015) Steam power plant configuration, design, and control. Wiley Interdiscip Rev: Energy Environ 6:537–563

Yoo H, Gajic Z (2018) New designs of linear observers and observer-based controllers for singularly perturbed linear systems. IEEE Trans Auto Control 63:3904–3911

Zhang W (2012) Quantitative process control theory (1st ed) CRC Press. https://doi.org/10.1201/b11313

Zhang Y, Naidu DS, Cai CX, Zou Y (2014) Singular perturbations and time scales in control theories and applications: an overview 2002–2012. Int J Info Syst Sci 9:1–36

# Chapter 5
# Two-Stage Multirate Feedback Control Designs

**Abstract** This chapter addresses the challenges of feedback control design for systems with slow and fast eigenvalue modes using a single sampling rate, which can lead to information loss or increased computations. The study introduces multirate state feedback controllers, demonstrating that for linear time-invariant systems with slow and fast eigenvalue modes, successful design can be achieved through two-stage multirate sampling. Two approaches are presented based on the sampling sequence, with proof that the feedback controls derived stabilize the full-order system. The chapter compares the design and implementation aspects of both approaches and demonstrates the proposed control's applicability through simulations.

**Keywords** Feedback control · Multirate sampling · Singular perturbation · Two-stage design · Two-time-scale system

## 5.1 Introduction

Three-stage feedback controllers are built in the continuous-time domain in the preceding chapters. However, feedback controls are nowadays implemented digitally because of the latest advancements in computing technology. The initial developments in this field presumed uniform sample rates for the conversion of analog signals into digital format and vice versa. Later, multirate sampling was found to be necessary due to an increase in computing burden in large-scale digital control systems (Jury 1967; Glasson 2008). Several methods for the multirate sampling-based eigenvalue assignment of discrete-time systems have been presented in the last few decades. The primary inspiration for these methods comes from Chammas and Leondes (1979); Araki and Hagiwara (1986); Hagiwara and Araki (1988). In Chammas and Leondes (1979), the input sampled at $\alpha/T_s$ Hz is supplied to the system to generate output. This output is then sampled at $1/T_s$ Hz to feed the controller. In this, $\alpha$ is larger than or equal to the controllability index of the system. Further, in Araki and Hagiwara (1986), it is shown that the inputs in Chammas and Leondes (1979) can be sampled at a rate slower than $\alpha/T_s$ Hz. While in Hagiwara and Araki (1988), the inputs are

 
R. Munje et al., *Multi-stage Feedback Control Design for Multi-time-scale System*, Springer Tracts in Electrical and Electronics Engineering,
https://doi.org/10.1007/978-981-97-4592-0_5

sampled at $1/T_s$ Hz and the outputs are sampled $\beta/T_s$ Hz, where $\beta$ is the observability index. By sampling inputs and outputs at different rates, these methods generally provide enough flexibility in eigenvalue assignment and eliminate the necessity for the deployment of an observer.

Recently, these techniques have also been applied to the singularly perturbed two-time-scale system (Zhang et al. 2014; Naidu et al. 1987; Tiwari et al. 2012), in which eigenvalues of the system are grouped as dominant (slow) eigenvalues and non-dominant (fast) eigenvalues. However, these techniques are less suitable for industrial applications. Also, in some cases, they result in feedback gains with larger values due to ill-conditioned system matrices arising because of the eigenvalue groupings. In such a situation, two-stage feedback control designs (Phillips 1980; Mahmoud 1983; Mahmoud et al. 1985; Radisavljevic-Gajic 2015a, b), employing single-rate sampling, help to alleviate system ill-conditioning issues. Nonetheless, it is worth mentioning that, the response of fast states is crucial during a short transient period. After that, the behavior of the system is commonly governed by slow states. The use of a slower sampling rate for feedback control design causes loss of information in fast-varying states. On the other hand, a faster sampling rate increases online computations considerably. Hence, the design of feedback control at a single sampling rate is impractical.

In this chapter, two multirate sampling approaches are proposed with the exact separation of slow and fast subsystems (Munje and Patre 2023). The designs of multirate feedback controllers (MFCs), presented in this chapter, are completed in two stages. The first technique discretizes the singularly perturbed continuous-time system with a smaller sample period after transforming it into an upper triangular form with a fully isolated fast subsystem. In contrast, the second technique discretizes the continuous-time system with a sample period larger than the previous strategy, obtaining the slow subsystem by first transforming it into a lower triangular form. The first stage of feedback control design is carried out in both schemes following block-triangularization, and the resulting discrete-time system is block-diagonalized to carry out the second step of feedback control design at a different sampling period. Based on the design, implementation, and simulation findings, these two approaches are examined.

## 5.2 Design Schemes

Consider a linear time-invariant, continuous-time system

$$\dot{\mathbf{z}}(t) = \mathbf{A}\mathbf{z}(t) + \mathbf{B}\mathbf{u}(t), \tag{5.1}$$

$$\mathbf{y}(t) = \mathbf{C}\mathbf{z}(t), \tag{5.2}$$

where $\mathbf{z} \in \Re^n$, $\mathbf{u} \in \Re^m$ and $\mathbf{y} \in \Re^p$ are the state, input and output variables respectively. If the system (5.1)–(5.2) is assumed to have two-time-scale structure, i.e., $n$

eigenvalues of the system have $n_1$ slow eigenvalues and $n_2$ fast eigenvalues, then it can be described by

$$\dot{\mathbf{z}}_1(t) = \mathbf{A}_{11}\mathbf{z}_1(t) + \mathbf{A}_{12}\mathbf{z}_2(t) + \mathbf{B}_1\mathbf{u}(t), \tag{5.3}$$

$$\varepsilon\dot{\mathbf{z}}_2(t) = \mathbf{A}_{21}\mathbf{z}_1(t) + \mathbf{A}_{22}\mathbf{z}_2(t) + \mathbf{B}_2\mathbf{u}(t), \tag{5.4}$$

$$\mathbf{y}(t) = \mathbf{C}_1\mathbf{z}_1(t) + \mathbf{C}_2\mathbf{z}_2(t), \tag{5.5}$$

where $\mathbf{z}_1 \in \Re^{n_1}$ and $\mathbf{z}_2 \in \Re^{n_2}$, such that $n_1 + n_2 = n$. Submatrices $\mathbf{A}_{ij}$, $\mathbf{B}_i$ and $\mathbf{C}_i$ are of suitable dimensions. The parameter $\varepsilon > 0$ denotes the speed ratio of the slow versus fast state variables. Before proceeding with the analysis, the following assumptions are made.

**Assumption 5.1** The pair $(\mathbf{A}, \mathbf{B})$ is controllable, where

$$\mathbf{A} = \begin{bmatrix} \mathbf{A}_{11} & \mathbf{A}_{12} \\ \frac{\mathbf{A}_{21}}{\varepsilon} & \frac{\mathbf{A}_{22}}{\varepsilon} \end{bmatrix} \text{ and } \mathbf{B} = \begin{bmatrix} \mathbf{B}_1 \\ \frac{\mathbf{B}_2}{\varepsilon} \end{bmatrix}. \tag{5.6}$$

**Assumption 5.2** $\mathbf{A}_{22}$ is invertible (Naidu 1988).

The state feedback control guarantees the stability of (5.3)–(5.4) according to Assumption 5.1. By using similarity transformations, it also aids in satisfying the controllability property of the lower-order continuous-time subsystems that are created in the next subsections. A common Assumption in singular perturbation theory, 5.2, is used to solve algebraic equations that are discussed in the ensuing subsections.

It makes sense to measure the slow subsystem state at a slower sampling rate than the fast subsystem state since the slow subsystem state is nearly constant throughout a time interval of $1/\varepsilon$. Since $\Delta = \tau/N$, where $N$ is the greatest integer smaller than or equal to $1/\varepsilon$, slow state variables can thus be sampled at the $\tau$ sampling interval if the fast state variables are sampled at the $\Delta$ sampling interval (Lennarston 1989). It is noteworthy that, under suitable sample period selection, the discrete-time system that is equal to the continuous two-time-scale system (5.3)–(5.4) would likewise exhibit the two-time-scale property (Tiwari et al. 2012; Naidu and Rao 1985). In order to ensure that $N$ is $\leq 1/\varepsilon$, the designer must choose among $N$, $\tau$, and $\Delta$. One can now obtain $\Delta$ if $\tau$ is chosen. With the two constraints listed below in mind, the sampling period $\tau$ can be carefully selected.

**Condition 5.1** *The sampling period ($T_s$) must be $\leq T_c/2$, where $T_c$ is the time period of the highest frequency (Jerri 1977).*

**Condition 5.2** *If Assumption 5.1 is satisfied and the system (5.3)–(5.4) has only real eigenvalues, then any $T_s > 0$ can be taken, otherwise*

$$T_s \neq \frac{2\pi m}{|\Im m(\psi_i - \psi_j)|}, \qquad \text{for } m = 1, 2, \ldots \tag{5.7}$$

*when $|\Re e(\psi_i - \psi_j)| = 0$, where $\psi_i$ and $\psi_j$ are any two eigenvalues (Chen 1999; Ding et al. 2009).*

Condition 5.1 must be satisfied first and then Condition 5.2. Condition 5.1 is well-known sampling theorem (Jerri 1977). Condition 5.2 is required to satisfy the controllability property of the discrete-time systems received by sampling the corresponding continuous-time system. It must be noted that, if $\tau$ meets Conditions 5.1 and 5.2, $\Delta$ too meets both.

Let us assume that the sampling starts at a moment $k\tau$. In this case, the fast state variables are measured at every $\Delta = k\tau/N$, whereas the slow state variables are measured at every $k\tau$, where $k = 0, 1, 2, \ldots$. In other words, fast state variables are sampled $N$ times at each $l\Delta$ in the $\tau$ sampling interval of slow state variables, where $l = 0, 1, 2, \ldots, (N-1)$. The overall feedback control input is the combination of slow and fast controls, such that

$$\mathbf{u}(t) = \mathbf{u}_s(k\tau) + \mathbf{u}_f(k\tau + l\Delta), \tag{5.8}$$

where $\mathbf{u}_s(k\tau)$ and $\mathbf{u}_f(k\tau + l\Delta)$ are piece-wise constant during

$$\begin{cases} \mathbf{u}_s(k\tau) & k\tau \le t < (k+1)\tau, \\ \mathbf{u}_f(k\tau + l\Delta) & k\tau + l\Delta \le t < k\tau + (l+1)\Delta, \end{cases}$$

respectively.

In most cases, the system does not need to explicitly include parameter $\varepsilon$. Even when a system is not in an explicit singular perturbation form, it can however contain slow and fast eigenvalues. Therefore, using singular perturbation techniques might not always be practical. Alternatively, a more promising approach would be to derive decoupled subsystems using similarity transformations (Ladde and Rajalakshmi 1985). This is why similarity transformations are used in the proposed two-stage multirate feedback control architecture. To decouple any one subsystem (either slow or fast), the original continuous-time system is first transformed into a block-triangular form. The next step is to discretize the block-triangular system. For the discretization of the block-triangular system, the sampling rate ($1/\Delta$ or $1/\tau$) is chosen according to the decoupled subsystem, i.e., a faster sampling rate is chosen if the fast subsystem is decoupled, and a slower sampling rate otherwise. Subsequently, the full-order discrete-time system is subjected to the designed subsystem control, marking the completion of the first step of feedback control design. The feedback control of the left subsystem is developed at a different sampling rate in the second stage. However, complete decoupling of the left subsystem is necessary first using a state variable transformation. The system must be described in an equivalent discrete-time form since the control is to be constructed at the alternate sampling rate. The second stage of design can then start. Precisely, the fast and slow subsystem controls are developed at distinct sampling rates through accurate subsystem separation in two design stages. As a result, multirate discretization modeling of the fast and slow subsystems is part of the proposed two-stage design. As previously mentioned, there are two approaches to the design of MFC (5.8), one is to sample fast subsystem states first, followed by slow subsystem states, and vice versa. The explanation of these two MFC designs is provided in the subsequent subsections.

### 5.2.1 MFC 1: Fast to Slow Sampling Scheme

In this scheme, the system (5.3)–(5.4) is changed to an upper triangular form

$$\begin{bmatrix} \dot{\mathbf{z}}_1(t) \\ \dot{\mathbf{z}}_{f_U}(t) \end{bmatrix} = \begin{bmatrix} \mathbf{A}_{s_U} & \mathbf{A}_{12} \\ \mathbf{0} & \frac{\mathbf{A}_{f_U}}{\varepsilon} \end{bmatrix} \begin{bmatrix} \mathbf{z}_1(t) \\ \mathbf{z}_{f_U}(t) \end{bmatrix} + \begin{bmatrix} \mathbf{B}_1 \\ \frac{\mathbf{B}_{f_U}}{\varepsilon} \end{bmatrix} \mathbf{u}(t), \tag{5.9}$$

such that $max|\psi(\mathbf{A}_{s_U})| \ll min\left|\psi\left(\frac{\mathbf{A}_{f_U}}{\varepsilon}\right)\right|$, where $\psi(.)$ represents eigenvalue and subscript 'U' denotes elements in the upper triangular form. The upper triangular form (5.9) is the outcome of the application of state variable transformation

$$\mathbf{z}_{f_U}(t) = \mathbf{P}_U \mathbf{z}_1(t) + \mathbf{z}_2(t), \tag{5.10}$$

to the system (5.3)–(5.4), in which $(n_2 \times n_1)$ matrix $\mathbf{P}_U$ satisfies

$$\varepsilon \mathbf{P}_U \mathbf{A}_{11} - \varepsilon \mathbf{P}_U \mathbf{A}_{12} \mathbf{P}_U + \mathbf{A}_{21} - \mathbf{A}_{22} \mathbf{P}_U = 0. \tag{5.11}$$

Unique solution of algebraic equation (5.11) for $\mathbf{P}_U$ can be found under Assumption 5.2, i.e., $det[\mathbf{A}_{22}] \neq 0$, for a sufficiently small value of $\varepsilon$ (Naidu 1988). Manipulating (5.3)–(5.4) and (5.10), one can get $\mathbf{A}_{s_U} = \mathbf{A}_{11} - \mathbf{A}_{12}\mathbf{P}_U$, $\mathbf{A}_{f_U} = \varepsilon \mathbf{P}_U \mathbf{A}_{12} + \mathbf{A}_{22}$ and $\mathbf{B}_{f_U} = \varepsilon \mathbf{P}_U \mathbf{B}_1 + \mathbf{B}_2$. In (5.9), the fast subsystem is entirely decoupled from the slow subsystem. At this point, as per Assumption 5.1, controllability of $\left(\frac{\mathbf{A}_{f_U}}{\varepsilon}, \frac{\mathbf{B}_{f_U}}{\varepsilon}\right)$ is preserved. Sampling the system (5.9) with $\Delta$ sampling interval yields

$$\begin{bmatrix} \mathbf{z}_{1,l+1} \\ \mathbf{z}_{f_U,l+1} \end{bmatrix} = \begin{bmatrix} \mathbf{\Phi}_{\Delta s_U} & \mathbf{\Phi}_{\Delta 12} \\ \mathbf{0} & \mathbf{\Phi}_{\Delta f_U} \end{bmatrix} \begin{bmatrix} \mathbf{z}_{1,l} \\ \mathbf{z}_{f_U,l} \end{bmatrix} + \begin{bmatrix} \mathbf{\Gamma}_{\Delta 1} \\ \mathbf{\Gamma}_{\Delta f_U} \end{bmatrix} \mathbf{u}_l, \tag{5.12}$$

where

$$\begin{bmatrix} \mathbf{\Phi}_{\Delta s_U} & \mathbf{\Phi}_{\Delta 12} \\ \mathbf{0} & \mathbf{\Phi}_{\Delta f_U} \end{bmatrix} = \mathbf{\Phi}_{\Delta_U} = e^{\bar{\mathbf{A}}_U \Delta}, \text{ and } \begin{bmatrix} \mathbf{\Gamma}_{\Delta 1} \\ \mathbf{\Gamma}_{\Delta f_U} \end{bmatrix} = \mathbf{\Gamma}_{\Delta_U} = \int_0^{\Delta} e^{\bar{\mathbf{A}}_U \delta} \bar{\mathbf{B}}_U d\delta,$$

in which

$$\bar{\mathbf{A}}_U = \begin{bmatrix} \mathbf{A}_{s_U} & \mathbf{A}_{12} \\ \mathbf{0} & \frac{\mathbf{A}_{f_U}}{\varepsilon} \end{bmatrix} \text{ and } \bar{\mathbf{B}}_U = \begin{bmatrix} \mathbf{B}_1 \\ \frac{\mathbf{B}_{f_U}}{\varepsilon} \end{bmatrix}.$$

Feedback control design for the discrete-time system (5.12) requires:

**Assumption 5.3** The system $(\mathbf{\Phi}_{\Delta_U}, \mathbf{\Gamma}_{\Delta_U})$ is controllable.

If sampling periods ($\tau$ and $\Delta$) are chosen so as to meet Conditions 5.1 and 5.2, Assumption 5.3 is satisfied. As a result, the controllability of $(\mathbf{\Phi}_{\Delta_U}, \mathbf{\Gamma}_{\Delta_U})$ implies the controllability of $(\mathbf{\Phi}_{\Delta f_U}, \mathbf{\Gamma}_{\Delta f_U})$. In the first stage, the feedback control signal

$$\mathbf{u}_l = -\mathbf{F}_{f_U}\mathbf{z}_{f_U,l} + \mathbf{u}_{s_U,l}, \tag{5.13}$$

is provided as an input to the system (5.12) as

$$\begin{bmatrix} \mathbf{z}_{1,l+1} \\ \mathbf{z}_{f_U,l+1} \end{bmatrix} = \begin{bmatrix} \mathbf{\Phi}_{\Delta s_U} & \mathbf{\Phi}_{\Delta 12} - \mathbf{\Gamma}_{\Delta 1}\mathbf{F}_{f_U} \\ \mathbf{0} & \mathbf{\Phi}_{\Delta f_U} - \mathbf{\Gamma}_{\Delta f_U}\mathbf{F}_{f_U} \end{bmatrix} \begin{bmatrix} \mathbf{z}_{1,l} \\ \mathbf{z}_{f_U,l} \end{bmatrix} + \begin{bmatrix} \mathbf{\Gamma}_{\Delta 1} \\ \mathbf{\Gamma}_{\Delta f_U} \end{bmatrix} \mathbf{u}_{s_U,l}, \tag{5.14}$$

where the feedback gain matrix $\mathbf{F}_{f_U}$ of dimension $(m \times n_2)$ is chosen, such that $|\psi(\mathbf{\Phi}_{\Delta f_U} - \mathbf{\Gamma}_{\Delta f_U}\mathbf{F}_{f_U})| < 1$ and the closed-loop eigenvalues are positioned near the origin. At this moment, using the change of states as

$$\mathbf{z}_{s_U,l} = \mathbf{z}_{1,l} - \mathbf{M}_U\mathbf{z}_{f_U,l}, \tag{5.15}$$

the system (5.14) is modified to

$$\begin{bmatrix} \mathbf{z}_{s_U,l+1} \\ \mathbf{z}_{f_U,l+1} \end{bmatrix} = \begin{bmatrix} \mathbf{\Phi}_{\Delta s_U} & \mathbf{0} \\ \mathbf{0} & \mathbf{\Phi}_{\Delta f_U} - \mathbf{\Gamma}_{\Delta f_U}\mathbf{F}_{f_U} \end{bmatrix} \begin{bmatrix} \mathbf{z}_{s_U,l} \\ \mathbf{z}_{f_U,l} \end{bmatrix} + \begin{bmatrix} \mathbf{\Gamma}_{\Delta s_U} \\ \mathbf{\Gamma}_{\Delta f_U} \end{bmatrix} \mathbf{u}_{s_U,l}, \tag{5.16}$$

where $\mathbf{\Gamma}_{\Delta s_U} = \mathbf{\Gamma}_{\Delta 1} - \mathbf{M}_U\mathbf{\Gamma}_{\Delta f_U}$. The matrix $\mathbf{M}_U$ of order $(n_1 \times n_2)$ is evaluated by setting

$$\mathbf{\Phi}_{\Delta 12} + \mathbf{\Phi}_{\Delta s_U}\mathbf{M}_U - \mathbf{\Gamma}_{\Delta 1}\mathbf{F}_{f_U} - \mathbf{M}_U(\mathbf{\Phi}_{\Delta f_U} - \mathbf{\Gamma}_{\Delta f_U}\mathbf{F}_{f_U}) = 0. \tag{5.17}$$

Equation (5.17) can be solved as a linear algebraic Sylvester equation (Chen 1999) using 'lyap' command of MATLAB (MATLAB/Simulink Control Design Toolbox User Manual 2017) with assumption:

**Assumption 5.4** $\psi(\mathbf{\Phi}_{\Delta s_U}) \neq \psi(\mathbf{\Phi}_{\Delta f_U} - \mathbf{\Gamma}_{\Delta f_U}\mathbf{F}_{f_U})$.

Assumption 5.4 is fulfilled because subsystem $(\mathbf{\Phi}_{\Delta f_U} - \mathbf{\Gamma}_{\Delta f_U}\mathbf{F}_{f_U})$ is user designed with the eigenvalues close to the origin, while the subsystem $\mathbf{\Phi}_{\Delta s_U}$ has original system's eigenvalues that are near to the unit circle. It is worth mentioning that, operating state transformation (5.15) on the system (5.14) maps it from triangular form to diagonal form (5.16). As a deduction, the controllability property of the uncoupled slow subsystem $(\mathbf{\Phi}_{\Delta s_U}, \mathbf{\Gamma}_{\Delta s_U})$ is conserved as per Assumption 5.3. The discrete-time system (5.16) has a sampling period of $\Delta$ seconds. If $\tau = \Delta N$, system corresponding to $\tau$ sampling interval can be acquired from (5.16) as

$$\begin{bmatrix} \mathbf{z}_{s_U,k+1} \\ \mathbf{z}_{f_U,k+1} \end{bmatrix} = \begin{bmatrix} \mathbf{\Phi}_{\tau s_U} & \mathbf{0} \\ \mathbf{0} & \Phi_{\tau f_U} \end{bmatrix} \begin{bmatrix} \mathbf{z}_{s_U,k} \\ \mathbf{z}_{f_U,k} \end{bmatrix} + \begin{bmatrix} \mathbf{\Gamma}_{\tau s_U} \\ \mathbf{\Gamma}_{\tau f_U} \end{bmatrix} \mathbf{u}_{s_U,k}, \tag{5.18}$$

where

$$\mathbf{\Phi}_{\tau s_U} = \mathbf{\Phi}_{\Delta s_U}^N, \tag{5.19}$$

$$\Phi_{\tau f_U} = (\mathbf{\Phi}_{\Delta f_U} - \mathbf{\Gamma}_{\Delta f_U}\mathbf{F}_{f_U})^N, \tag{5.20}$$

$$\boldsymbol{\Gamma}_{\tau s_U} = \sum_{i=0}^{N-1} \boldsymbol{\Phi}^i_{\Delta s_U} \boldsymbol{\Gamma}_{\Delta s_U}, \tag{5.21}$$

$$\boldsymbol{\Gamma}_{\tau f_U} = \sum_{i=0}^{N-1} (\boldsymbol{\Phi}_{\Delta f_U} - \boldsymbol{\Gamma}_{\Delta f_U}\mathbf{F}_{f_U})^i \boldsymbol{\Gamma}_{\Delta f_U}. \tag{5.22}$$

Herein, the controllability of the slow subsystem $(\boldsymbol{\Phi}_{\tau s_U}, \boldsymbol{\Gamma}_{\tau s_U})$ is confirmed by Assumption 5.3. Now in the second stage, local feedback control

$$\mathbf{u}_{s_U,k} = -\mathbf{F}_{s_U}\mathbf{z}_{s_U,k}, \tag{5.23}$$

is designed and passed to the system (5.18). The input (5.23) can also be treated as $\mathbf{u}_{s_U,l} = -\mathbf{F}_{s_U}\mathbf{z}_{s_U,l}$, where $\mathbf{z}_{s_U,l}$ is constant for one complete $\tau$ sampling interval. After applying (5.23) to (5.18), a closed-loop system

$$\begin{bmatrix} \mathbf{z}_{s_U,k+1} \\ \mathbf{z}_{f_U,k+1} \end{bmatrix} = \begin{bmatrix} \boldsymbol{\Phi}_{\tau s_U} - \boldsymbol{\Gamma}_{\tau s_U}\mathbf{F}_{s_U} & \mathbf{0} \\ -\boldsymbol{\Gamma}_{\tau f_U}\mathbf{F}_{s_U} & \Phi_{\tau f_U} \end{bmatrix} \begin{bmatrix} \mathbf{z}_{s_U,k} \\ \mathbf{z}_{f_U,k} \end{bmatrix}, \tag{5.24}$$

is obtained, where $\mathbf{F}_{s_U}$ is a feedback gain matrix of order $(m \times n_1)$, selected, such that $|\psi(\boldsymbol{\Phi}_{\tau s_U} - \boldsymbol{\Gamma}_{\tau s_U}\mathbf{F}_{s_U})| < 1$ and placed near the unit circle. Accordingly, the overall multirate state feedback control input is written as

$$\mathbf{u} = -\mathbf{F}_{f_U}\mathbf{z}_{f_U,l} - \mathbf{F}_{s_U}\mathbf{z}_{s_U,k}, \tag{5.25}$$

where

$$\begin{aligned} \mathbf{u}_f(k\tau + l\Delta) &= -\mathbf{F}_{f_U}\mathbf{z}_{f_U,l}, \quad k\tau + l\Delta \le t < k\tau + (l+1)\Delta \\ \mathbf{u}_s(k\tau) &= -\mathbf{F}_{s_U}\mathbf{z}_{s_U,k}, \qquad k\tau \le t < (k+1)\tau. \end{aligned}$$

**Theorem 5.1** *The control input (5.25), received by sampling fast and slow subsystem states at $\Delta$ and $\tau$ sampling intervals respectively, stabilizes the system (5.24).*

***Proof*** In a block-triangular system, eigenvalues are the disjoint sum of eigenvalues of the diagonal matrices. Hence, in (5.24) eigenvalues of $(\boldsymbol{\Phi}_{\tau s_U} - \boldsymbol{\Gamma}_{\tau s_U}\mathbf{F}_{s_U})$ and $\Phi_{\tau f_U}$ affirm the stability of the system. The subsystem $(\boldsymbol{\Phi}_{\tau s_U} - \boldsymbol{\Gamma}_{\tau s_U}\mathbf{F}_{s_U})$ is stabilized with feedback gain $\mathbf{F}_{s_U}$, such that the eigenvalues of it stay within the unit circle near perimeter. On the other hand, subsystem $\Phi_{\tau f_U}$ is given by (5.20), in which the eigenvalues of $(\boldsymbol{\Phi}_{\Delta f_U} - \boldsymbol{\Gamma}_{\Delta f_U}\mathbf{F}_{f_U})$ are stabilized by the feedback gain $\mathbf{F}_{f_U}$. At this instant, if

$$|\psi(\boldsymbol{\Phi}_{\Delta f_U} - \boldsymbol{\Gamma}_{\Delta f_U}\mathbf{F}_{f_U})| < 1, \tag{5.26}$$

then it is easy to prove that

$$|\psi((\mathbf{\Phi}_{\Delta f_U} - \mathbf{\Gamma}_{\Delta f_U}\mathbf{F}_{f_U})^N)| < 1. \tag{5.27}$$

Thus, $|\psi(\Phi_{\tau f_U})| < 1$. Hence, the control input (5.25) stabilizes the system (5.24). □

### 5.2.2 MFC 2: Slow to Fast Sampling Scheme

In this approach, the lower triangular form of the system (5.3)–(5.4) is built

$$\begin{bmatrix} \dot{\mathbf{z}}_{s_L}(t) \\ \dot{\mathbf{z}}_2(t) \end{bmatrix} = \begin{bmatrix} \mathbf{A}_{s_L} & \mathbf{0} \\ \frac{\mathbf{A}_{21}}{\varepsilon} & \frac{\mathbf{A}_{f_L}}{\varepsilon} \end{bmatrix} \begin{bmatrix} \mathbf{z}_{s_L}(t) \\ \mathbf{z}_2(t) \end{bmatrix} + \begin{bmatrix} \mathbf{B}_{s_L} \\ \frac{\mathbf{B}_2}{\varepsilon} \end{bmatrix} \mathbf{u}(t), \tag{5.28}$$

so that, $max|\psi(\mathbf{A}_{s_L})| \ll min\left|\psi\left(\frac{\mathbf{A}_{f_L}}{\varepsilon}\right)\right|$, where the subscript 'L' indicates elements in the lower triangular form. It is important to note that, despite differences in $\mathbf{A}_{s_U}$ and $\mathbf{A}_{s_L}$, and $\frac{\mathbf{A}_{f_U}}{\varepsilon}$ and $\frac{\mathbf{A}_{f_L}}{\varepsilon}$, the slow eigenvalues of $\mathbf{A}_{s_U}$ and $\mathbf{A}_{s_L}$ and fast eigenvalues of $\frac{\mathbf{A}_{f_U}}{\varepsilon}$ and $\frac{\mathbf{A}_{f_L}}{\varepsilon}$ remain same. The linear transformation employed to attain a lower triangular form (5.28) is

$$\mathbf{z}_{s_L}(t) = \mathbf{z}_1(t) + \varepsilon \mathbf{P}_L \mathbf{z}_2(t), \tag{5.29}$$

in which $\mathbf{P}_L$ of matrix size $(n_1 \times n_2)$ is determined by solving

$$-\varepsilon(\mathbf{A}_{11} + \mathbf{P}_L\mathbf{A}_{21})\mathbf{P}_L + \mathbf{A}_{12} + \mathbf{P}_L\mathbf{A}_{22} = 0, \tag{5.30}$$

iteratively for a small value of $\varepsilon$ (Naidu 1988) with Assumption 5.2. Substituting (5.29) in (5.3)–(5.4) one can easily gain $\mathbf{A}_{s_L} = \mathbf{A}_{11} + \mathbf{P}_L\mathbf{A}_{21}$, $\mathbf{A}_{f_L} = -\varepsilon\mathbf{A}_{21}\mathbf{P}_L + \mathbf{A}_{22}$ and $\mathbf{B}_{s_L} = \mathbf{B}_1 + \mathbf{P}_L\mathbf{B}_2$. Here, the slow subsystem is totally decoupled from the fast subsystem, but it holds controllability property due to Assumption 5.1. Sampling the system (5.28) at the rate of $1/\tau$ generates an equivalent discrete-time system

$$\begin{bmatrix} \mathbf{z}_{s_L,k+1} \\ \mathbf{z}_{2,k+1} \end{bmatrix} = \begin{bmatrix} \mathbf{\Phi}_{\tau s_L} & \mathbf{0} \\ \mathbf{\Phi}_{\tau 21} & \mathbf{\Phi}_{\tau f_L} \end{bmatrix} \begin{bmatrix} \mathbf{z}_{s_L,k} \\ \mathbf{z}_{2,k} \end{bmatrix} + \begin{bmatrix} \mathbf{\Gamma}_{\tau s_L} \\ \mathbf{\Gamma}_{\tau 2} \end{bmatrix} \mathbf{u}_k, \tag{5.31}$$

where

$$\begin{bmatrix} \mathbf{\Phi}_{\tau s_L} & \mathbf{0} \\ \mathbf{\Phi}_{\tau 21} & \mathbf{\Phi}_{\tau f_L} \end{bmatrix} = \mathbf{\Phi}_{\tau_L} = e^{\bar{\mathbf{A}}_L \tau}, \text{ and } \begin{bmatrix} \mathbf{\Gamma}_{\tau s_L} \\ \mathbf{\Gamma}_{\tau 2} \end{bmatrix} = \mathbf{\Gamma}_{\tau_L} = \int_0^{\tau} e^{\bar{\mathbf{A}}_L \omega} \bar{\mathbf{B}}_L d\omega,$$

with

$$\bar{\mathbf{A}}_L = \begin{bmatrix} \mathbf{A}_{s_L} & \mathbf{0} \\ \frac{\mathbf{A}_{21}}{\varepsilon} & \frac{\mathbf{A}_{f_L}}{\varepsilon} \end{bmatrix} \text{ and } \bar{\mathbf{B}}_L = \begin{bmatrix} \mathbf{B}_{s_L} \\ \frac{\mathbf{B}_2}{\varepsilon} \end{bmatrix}.$$

Designing feedback control for the system (5.31) requires the next assumption.

**Assumption 5.5** The system $(\mathbf{\Phi}_{\tau_L}, \mathbf{\Gamma}_{\tau_L})$ is controllable.

Once again, Assumption 5.5 is satisfied if sampling periods ($\tau$ and $\Delta$) are selected referring to Conditions 5.1 and 5.2. The slow subsystem $(\mathbf{\Phi}_{\tau s_L}, \mathbf{\Gamma}_{\tau s_L})$ is, thus controllable due to Assumption 5.5. Employing feedback control, in the first stage as

$$\mathbf{u}_k = -\mathbf{F}_{s_L}\mathbf{z}_{s_L,k} + \mathbf{u}_{f,k}. \tag{5.32}$$

to the system (5.31) gives

$$\begin{bmatrix} \mathbf{z}_{s_L,k+1} \\ \mathbf{z}_{2,k+1} \end{bmatrix} = \begin{bmatrix} \mathbf{\Phi}_{\tau s_L} - \mathbf{\Gamma}_{\tau s_L}\mathbf{F}_{s_L} & \mathbf{0} \\ \mathbf{\Phi}_{\tau 21} - \mathbf{\Gamma}_{\tau 2}\mathbf{F}_{s_L} & \mathbf{\Phi}_{\tau f_L} \end{bmatrix} \begin{bmatrix} \mathbf{z}_{s_L,k} \\ \mathbf{z}_{2,k} \end{bmatrix} + \begin{bmatrix} \mathbf{\Gamma}_{\tau s_L} \\ \mathbf{\Gamma}_{\tau 2} \end{bmatrix} \mathbf{u}_{f,k}, \tag{5.33}$$

where feedback gain matrix $\mathbf{F}_{s_L}$ of dimensionality $(m \times n_1)$ is appropriately calculated, such that the eigenvalues of $(\mathbf{\Phi}_{\tau s_L} - \mathbf{\Gamma}_{\tau s_L}\mathbf{F}_{s_L})$ are asymptotically stable and are located near the circumference of the unit circle. After that, introducing a change of coordinates as

$$\mathbf{z}_{f_L,k} = -\mathbf{M}_L\mathbf{z}_{s_L,k} + \mathbf{z}_{2,k}, \tag{5.34}$$

the system (5.33) is structured into diagonal form

$$\begin{bmatrix} \mathbf{z}_{s_L,k+1} \\ \mathbf{z}_{f_L,k+1} \end{bmatrix} = \begin{bmatrix} \mathbf{\Phi}_{\tau s_L} - \mathbf{\Gamma}_{\tau s_L}\mathbf{F}_{s_L} & \mathbf{0} \\ \mathbf{0} & \mathbf{\Phi}_{\tau f_L} \end{bmatrix} \begin{bmatrix} \mathbf{z}_{s_L,k} \\ \mathbf{z}_{f_L,k} \end{bmatrix} + \begin{bmatrix} \mathbf{\Gamma}_{\tau s_L} \\ \mathbf{\Gamma}_{\tau f_L} \end{bmatrix} \mathbf{u}_{f_L,k}, \tag{5.35}$$

where $\mathbf{\Gamma}_{\tau f_L} = -\mathbf{M}_L\mathbf{\Gamma}_{\tau s_L} + \mathbf{\Gamma}_{\tau 2}$ and $(n_2 \times n_1)$ matrix $\mathbf{M}_L$ is computed by exercising

$$\mathbf{\Phi}_{\tau 21} - \mathbf{M}_L(\mathbf{\Phi}_{\tau s_L} - \mathbf{\Gamma}_{\tau s_L}\mathbf{F}_{s_L}) + \mathbf{\Phi}_{\tau f_L}\mathbf{M}_L - \mathbf{\Gamma}_{\tau 2}\mathbf{F}_{s_L} = 0. \tag{5.36}$$

Equation (5.36) can also be solved as a linear algebraic Sylvester equation (Chen 1999), similar to (5.17) with

**Assumption 5.6** $\psi(\mathbf{\Phi}_{\tau s_L} - \mathbf{\Gamma}_{\tau s_L}\mathbf{F}_{s_L}) \neq \psi(\mathbf{\Phi}_{\tau f_L})$.

It can be convinced by the fact, that the closed-loop subsystem $(\mathbf{\Phi}_{\tau s_L} - \mathbf{\Gamma}_{\tau s_L}\mathbf{F}_{s_L})$ is designed with desired eigenvalues situated within unit circle near the periphery and the subsystem $(\mathbf{\Phi}_{\tau f_L})$ has original system's eigenvalues distributed around the origin. Further, the controllability of decoupled subsystem $(\mathbf{\Phi}_{\tau f_L}, \mathbf{\Gamma}_{\tau f_L})$ is guaranteed by Assumption 5.5. The discrete-time system (5.35) has a sampling period of $\tau$ seconds. In this, the fast subsystem can be rewritten as

$$\mathbf{z}_{f_L,k+1} = \mathbf{\Phi}_{\tau f_L}\mathbf{z}_{f_L,k} + \mathbf{\Gamma}_{\tau f_L}\mathbf{u}_{f_L,k}. \tag{5.37}$$

An equivalent system of (5.37) sampled as $\Delta$ sampling interval can be constructed using 'd2c' and 'c2d' commands of MATLAB (MATLAB/Simulink Control Design Toolbox User Manual 2017) as

$$\mathbf{z}_{f_L,l+1} = \mathbf{\Phi}_{\Delta f_L}\mathbf{z}_{f_L,l} + \mathbf{\Gamma}_{\Delta f_L}\mathbf{u}_{f_L,l}. \tag{5.38}$$

Subsystems (5.37) and (5.38) are mathematically related as

$$\mathbf{\Phi}_{\tau f_L} = \mathbf{\Phi}^N_{\Delta f_L} \text{ and } \mathbf{\Gamma}_{\tau f_L} = \sum_{i=0}^{N-1} \mathbf{\Phi}^i_{\Delta f_L}\mathbf{\Gamma}_{\Delta f_L}.$$

In the second stage, formulating a feedback control input

$$\mathbf{u}_{f_L,l} = -\mathbf{F}_{f_L}\mathbf{z}_{f,l}, \tag{5.39}$$

and passing it to the system (5.38) produces a closed-loop system as

$$\mathbf{z}_{f_L,l+1} = (\mathbf{\Phi}_{\Delta f_L} - \mathbf{\Gamma}_{\Delta f_L}\mathbf{F}_{f_L})\mathbf{z}_{f_L,l} \tag{5.40}$$

where $\mathbf{F}_{f_L}$ is $(m \times n_1)$ feedback gain matrix, such that $|\psi(\mathbf{\Phi}_{\Delta f_L} - \mathbf{\Gamma}_{\Delta f_L}\mathbf{F}_{f_L})| < 1$ and eigenvalues are close to the origin. The closed-loop system equivalent to (5.40) obtained for $\tau$ sampling interval can be reconstructed as

$$\mathbf{z}_{f_L,k+1} = (\mathbf{\Phi}_{\Delta f_L} - \mathbf{\Gamma}_{\Delta f_L}\mathbf{F}_{f_L})^N\mathbf{z}_{f_L,k}. \tag{5.41}$$

The resulting closed-loop system is given by

$$\begin{bmatrix}\mathbf{z}_{s_L,k+1}\\ \mathbf{z}_{f_L,k+1}\end{bmatrix} = \begin{bmatrix}\mathbf{\Phi}_{\tau s_L} - \mathbf{\Gamma}_{\tau s_L}\mathbf{F}_{s_L} & -\mathbf{\Gamma}_{\tau s_L}\bar{\mathbf{F}}_{f_L}\\ \mathbf{0} & (\mathbf{\Phi}_{\Delta f_L} - \mathbf{\Gamma}_{\Delta f_L}\mathbf{F}_{f_L})^N\end{bmatrix}\begin{bmatrix}\mathbf{z}_{s_L,k}\\ \mathbf{z}_{f_L,k}\end{bmatrix}, \tag{5.42}$$

where $\bar{\mathbf{F}}_{f_L}$ represents state feedback gain designed at $\tau$ sampling interval. Eventually, the global MFC input is

$$\mathbf{u} = -\mathbf{F}_{s_L}\mathbf{z}_{s_L,k} - \mathbf{F}_{f_L}\mathbf{z}_{f,l}, \tag{5.43}$$

where

$$\begin{aligned}\mathbf{u}_s(k\tau) &= -\mathbf{F}_{s_L}\mathbf{z}_{s_L,k}, \qquad k\tau \le t < (k+1)\tau,\\ \mathbf{u}_f(k\tau + l\Delta) &= -\mathbf{F}_{f_L}\mathbf{z}_{f_L,l}, \quad k\tau + l\Delta \le t < k\tau + (l+1)\Delta.\end{aligned}$$

**Theorem 5.2** *The MFC (5.43), designed by sampling slow and fast subsystem states respectively at $\tau$ and $\Delta$ sampling periods, achieves stabilization of the system (5.42).*

***Proof*** The control input (5.43) has two components. The first component signifies feedback gain calculated for the slow subsystem, such that eigenvalues of ($\mathbf{\Phi}_{\tau s_L} - \mathbf{\Gamma}_{\tau s_L}\mathbf{F}_{s_L}$) are made asymptotically stable by assigning them within and close to the unit circle in z-plane. The second component symbolizes feedback gain designed for the fast subsystem, so that,

$$|\psi(\mathbf{\Phi}_{\Delta f_L} - \mathbf{\Gamma}_{\Delta f_L}\mathbf{F}_{f_L})| < 1. \tag{5.44}$$

And, these fast eigenvalues are put around the origin. If (5.44) is satisfied, then

$$|\psi((\mathbf{\Phi}_{\Delta f_L} - \mathbf{\Gamma}_{\Delta f_L}\mathbf{F}_{f_L})^N)| < 1, \tag{5.45}$$

is also satisfied. This proves that the diagonal elements in (5.42) have asymptotically stable eigenvalues, making the entire system stable. □

**Remark 5.1** If the eigenvalues of fast subsystem $\frac{\mathbf{A}_{f_L}}{\varepsilon}$ are stable with reasonable stability properties, then feedback gain designed for the fast subsystem can be taken as zero, i.e., $\mathbf{F}_{f_L} = 0$. Thereof, the MFC (5.43) reduces to $\mathbf{u} = -\mathbf{F}_{s_L}\mathbf{z}_{s_L,k}$.

Novelty in the multirate feedback control design and resulting benefits are briefed as follows.

1. The design complexity is decreased because, in each design, only two state variable transformations as compared to three transformations in Phillips (1980) are required to finish a two-stage design.
2. Compared to three equations in Phillips (1980), the presented designs require solutions to only two algebraic equations, which reduces offline computations.
3. The design approaches that are being given employ a multirate sampling idea, which leads to two benefits: (i) a reduction in online calculations when compared to faster single-rate sampling (Radisavljevic-Gajic 2015a), and (ii) a significant improvement in closed-loop performance when compared to slower single-rate sampling (Radisavljevic-Gajic 2015b).
4. The designs are completed in two separate phases using similarity transformations, so any errors made in the second stage would not have an impact on the accuracy of the first-stage designs.

At last, the claimed multirate feedback controls are compared based on design procedure and implementation. The step-by-step design process is stated in Table 5.1, and its implementation is portrayed in Fig. 5.1.

**Table 5.1** Design comparison of multirate feedback controls

| Step | MFC 1 | MFC 2 |
|---|---|---|
| 1 | Solve (5.11) to get $\mathbf{P}_U$ with $\mathbf{P}_{U_{i+1}} = \mathbf{A}_{22}^{-1}(\varepsilon \mathbf{P}_{U_i}\mathbf{A}_{11} - \varepsilon \mathbf{P}_{U_i}\mathbf{A}_{12}\mathbf{P}_{U_i} + \mathbf{A}_{21})$, where $\mathbf{P}_{U_0} = \mathbf{A}_{22}^{-1}\mathbf{A}_{21}$ | Evaluate (5.30) for $\mathbf{P}_L$ with $\mathbf{P}_{L_{i+1}} = (\varepsilon(\mathbf{A}_{11} + \mathbf{P}_{L_i}\mathbf{A}_{21})\mathbf{P}_{L_i} - \mathbf{A}_{12})\mathbf{A}_{22}^{-1}$, where $\mathbf{P}_{L_0} = -\mathbf{A}_{12}\mathbf{A}_{22}^{-1}$ |
| 2 | Determine $\mathbf{A}_{s_U}$, $\mathbf{A}_{f_U}$, $\mathbf{B}_{s_U}$ and construct triangular form (5.9) | Calculate $\mathbf{A}_{s_L}$, $\mathbf{A}_{f_L}$, $\mathbf{B}_{f_L}$ and achieve triangular form (5.28) |
| 3 | Discretize system (5.9) with $\Delta$ sampling interval to get (5.12) | Discretize system (5.28) with $\tau$ sampling interval to get (5.31) |
| 4 | Obtain $\mathbf{F}_{f_U}$, such that $\psi(\mathbf{\Phi}_{\Delta f_U} - \mathbf{\Gamma}_{\Delta f_U}\mathbf{F}_{f_U}) = \psi_{\Delta f}^{desired}$ | Find out $\mathbf{F}_{s_L}$, such that $\psi(\mathbf{\Phi}_{\tau s_L} - \mathbf{\Gamma}_{\tau s_L}\mathbf{F}_{s_L}) = \psi_{\tau s}^{desired}$ |
| 5 | Compute $\mathbf{M}_U$ by solving (5.17) and determine $\mathbf{\Gamma}_{\Delta s_U}$ | Estimate $\mathbf{M}_L$ by exercising (5.36) and get $\mathbf{\Gamma}_{\tau f_L}$ |
| 6 | Build the system (5.18) with $\tau$ sampling interval using (5.19)–(5.22) | Get the system (5.37) with $\Delta$ sampling interval |
| 7 | Find out $\mathbf{F}_{s_U}$, such that $\psi(\mathbf{\Phi}_{\tau s_U} - \mathbf{\Gamma}_{\tau s_U}\mathbf{F}_{s_U}) = \psi_{\tau s}^{desired}$ | Obtain $\mathbf{F}_{f_L}$, such that $\psi(\mathbf{\Phi}_{\Delta f_L} - \mathbf{\Gamma}_{\Delta f_L}\mathbf{F}_{f_L}) = \psi_{\Delta f}^{desired}$ |

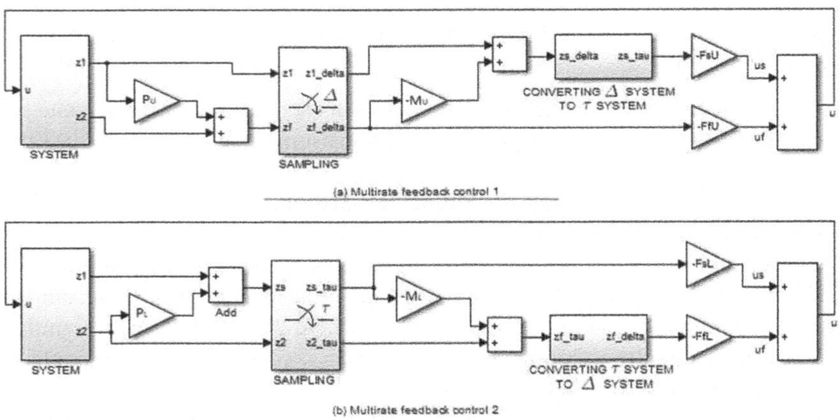


**Fig. 5.1** Implementation of the multirate feedback controls

## 5.3 Numerical Example

A fifth-order steam power system (Radisavljevic-Gajic 2015a), having two slow eigenvalues and three fast eigenvalues, has state and input matrices as

$$\mathbf{A} = \begin{bmatrix} -0.1125 & 0.0468 & 0.1049 & 0.0001 & 0.1876 \\ -0.0001 & -0.1669 & -0.0000 & -0.0002 & 0.6264 \\ -0.0000 & 1.3333 & -2.0001 & -0.0000 & -0.0004 \\ -2.0000 & 0.0001 & -0.0001 & -2.0001 & 0.0001 \\ 0.0006 & -0.0001 & 0.0007 & 1.2677 & -5.0014 \end{bmatrix},$$

$$\mathbf{B} = \begin{bmatrix} 0.0001 \; 0.1499 \; 0.0003 \; 1.5003 \; -0.0022 \end{bmatrix}^T.$$

Eigenvalues are located at $(-5.0332, \; -1.9697 \pm j0.1428, \; -0.1542 \pm j0.1494)$ with the ratio of slow versus fast states as

$$\varepsilon = \frac{max|\psi_s|}{min|\psi_f|} = 0.1087.$$

Thus, $N$ is taken as 9 ($< 1/\varepsilon$). If the sampling duration for the slow subsystem is selected as $\tau = 1.8$ s, then the fast subsystem states are sampled at every $\Delta = 0.2$ s. Supposing the desired eigenvalues of the continuous-time system be $(-0.2, -0.35, -3.1, -4.3, -5.5)$. Consequently, the desired eigenvalues in $s$-plane for the slow and fast subsystems for $\tau$ and $\Delta$ sampling intervals respectively can be given by $e^{\psi_i T_s}$, where $\psi_i$ is the $i$th eigenvalue of the continuous-time system and $T_s$ is the sampling period. Accordingly, one can obtain

$$\psi_{\tau s}^{desired} = (0.6977, 0.5326),$$
$$\psi_{\Delta f}^{desired} = (0.5379, 0.4232, 0.3329).$$

Using the fixed point iteration method (Naidu 1988), the solution to (5.11) is found in the first approach, leading to the upper triangular form (5.6). After that, this system is discretized with a 0.2 s sample period. After that, the feedback gain is calculated as given below, to place $\psi_{\Delta f}^{desired}$,

$$\mathbf{F}_{f_U} = \begin{bmatrix} -5.7277 \; 1.1959 \; 0.8986 \end{bmatrix}.$$

Next, using (5.19)–(5.22), the matrix $\mathbf{M}_U$ is estimated to obtain the diagonal system (5.16). It is then represented in the system (5.18) with $\tau$ sampling interval. The eigenvalues of the slow subsystem are placed at $\psi_{\tau s}^{desired}$ through the computation of the feedback gain

$$\mathbf{F}_{s_U} = \begin{bmatrix} -0.8079 \; 1.6862 \end{bmatrix}.$$

Using an iterative procedure, in the second approach, the matrix $\mathbf{P}_L$ is determined in (5.30), leading to the determination of a lower triangular form (5.28). Following that, an analogous discrete-time system (5.31) is achieved, featuring a sampling period of 1.8 s. The matching feedback gain is then obtained as follows if $\psi_{\tau s}^{desired}$ is used as the desired eigenvalues of the slow subsystem.

$$\mathbf{F}_{s_L} = \left[-0.2675\ 0.5814\right].$$

Using the 'lyap' command of MATLAB, Eq. (5.36) is solved and the decoupled fast subsystem (5.35) is obtained. Its continuous-time equivalent is obtained with the 'd2c' command, and it is discretized with the $\Delta$ sampling interval once again. Assuming $\psi_{\Delta f}^{desired}$ for the desired eigenvalues, the fast subsystem feedback gain is expressed as

$$\mathbf{F}_{f_L} = \left[-3.7176\ 0.4713\ 0.1945\right].$$

Utilizing $\mathbf{F}_{f_U}$ and $\mathbf{F}_{s_U}$ in the first scheme, control input (5.25) is formulated and similarly using $\mathbf{F}_{s_L}$ and $\mathbf{F}_{f_L}$ in the second scheme, control signal (5.43) is constructed. With these controls, the system is simulated with the initial condition $(-0.2, 0.1, 0.1, -0.2, 0.4)$. The simulation results are compared with the single-rate sampling viz. $\Delta$ and $\tau$ seconds. The slow and fast subsystem states' evolution is depicted in Figs. 5.2 and 5.3, respectively, while the corresponding control inputs are

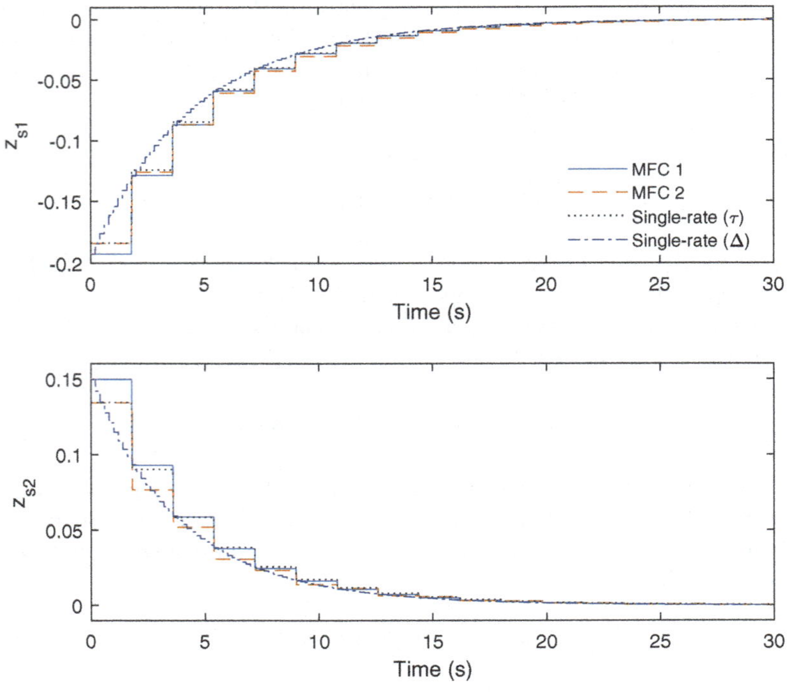


**Fig. 5.2** Evolution of slow subsystem states

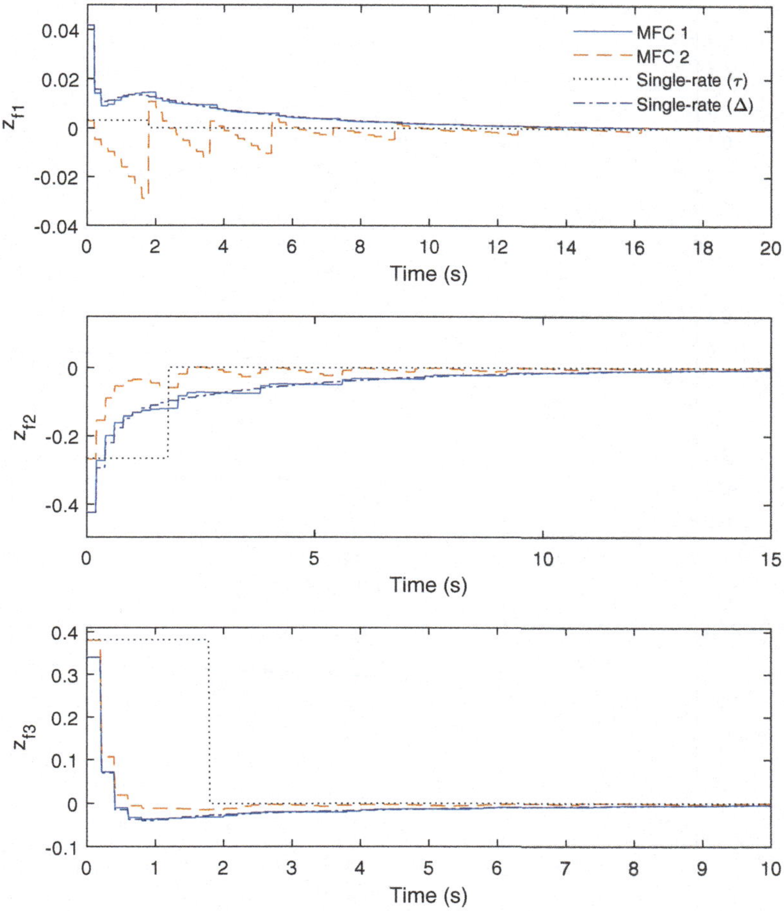


**Fig. 5.3** Evolution of fast subsystem states

shown in Fig. 5.4. As can be shown in Fig. 5.2, slow subsystem state fluctuations are going to be good with a single sampling rate $\Delta$. However, when multirate sampling is to be used, MFC 1 performs satisfactorily when compared to MFC 2 and single-rate sampling with $\tau$.

As seen in Fig. 5.3, even though the fast states reach steady-state in just two to three steps of single-rate sampling ($\tau$), they are unable to record the first transients that other controls are able to capture. MFC 1 responds smoothly in MFCs. However, MFC 2 produces oscillations for a single sample of $\tau$ seconds. This is because there are fewer samples overall and no oscillations or variations are seen when a system is changed from faster to slower sampling rates. Conversely, when the opposite is done, more samples are used, and as a result of discretization and reconstruction, certain oscillations are produced in the fast subsystem states (Satyanarayana and

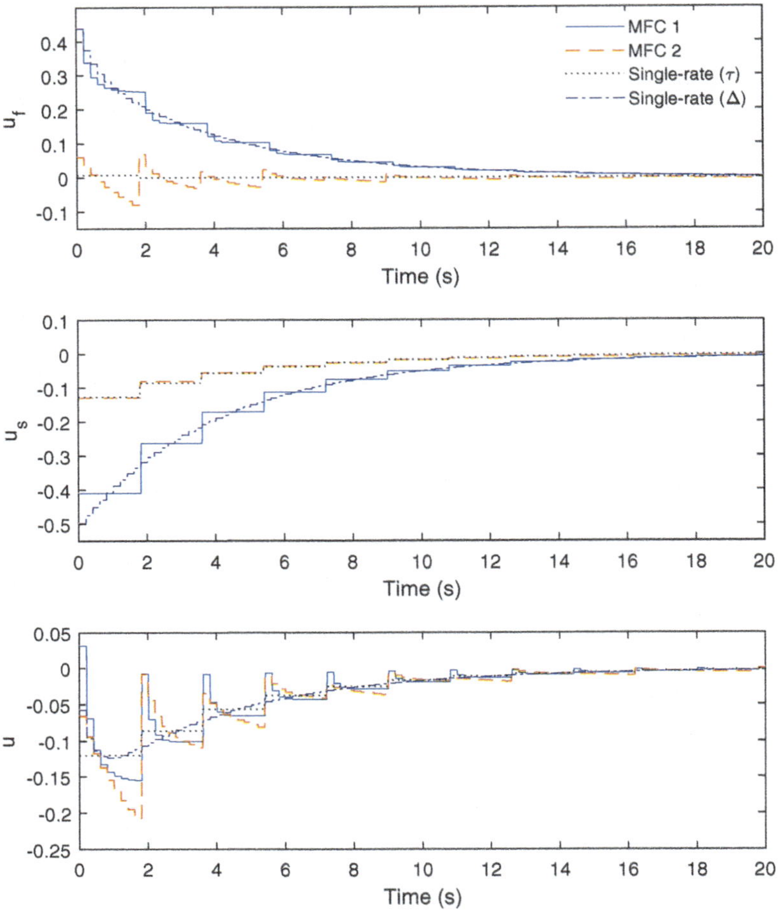


**Fig. 5.4** Multirate feedback control inputs

Janardhanan 2014; Priya and Bandyopadhyay 2017). The control input, which is a function of fast states, fluctuates in MFC 2 (Fig. 5.4) and takes longer to settle, even though the aggregate input due to MFC 1 and MFC 2 is the same. In conclusion, MFC 1 performs better than the other controls.

## 5.4 Conclusion

This chapter uses the exact separation of the slow and fast subsystems to offer multirate state feedback controllers for a two-time-scale system. Compared to the slow subsystem states, the fast subsystem states are sampled faster. Subsystem separation in two steps and multirate feedback control design in two stages are combined. First, a block-triangularized continuous-time singly perturbed system is discretized at a slower/faster sampling rate and split into slow/fast subsystems. The isolated subsystem is then developed with feedback control in mind. Second, the resulting system is represented in appropriate discrete-time form and block-diagonalized to decouple the fast/slow subsystem. Afterward, the fast/slow subsystem's feedback control is acquired at a faster/lower sampling rate. An example is given with these controls applied, and simulations are shown.

## References


Araki M, Hagiwara T (1986) Pole assignment by multirate sampled-data output feedback. Int J Control 44:1661–1673

Chammas AB, Leondes CT (1979) Pole assignment by piecewise constant output feedback. Int J Control 29:31–38

Chen CT (1999) Linear system theory and design. Oxford University Press, New York

Ding F, Qiu L, Chen T (2009) Reconstruction of continuous-time systems from their non-uniformly sampled discrete-time systems. Automatica 45:324–332

Glasson D (2008) Development and applications of multirate digital control [25 Years Ago]. IEEE Control Syst Mag 28:13–15

Hagiwara T, Araki M (1988) Design of a stable state feedback controller based on the multirate sampling of the plant output. IEEE Trans Auto Control 33:812–819

Jerri AJ (1977) The Shannon sampling theorem-its various extensions and applications: a tutorial review. Proc IEEE 65:1565–1596

Jury E (1967) A note on multirate sampled-data systems. IEEE Trans Autom Control 12:319–320

Ladde GS, Rajalakshmi SG (1985) Diagonalization and stability of multi-time- scale singularly perturbed linear systems. Appl Math Comp 16:115–140

Lennarston B (1989) Multirate sampled-data control of two-time-scale systems. IEEE Trans Auto Control 34:642–644

Mahmoud MS (1983) Design of feedback controllers by two-stage methods. Appl Math Model 7:163–168

Mahmoud MS, Chen T, Singh MG (1985) On eigenvalue assignment in discrete systems with fast and slow modes. Int J Syst Sci 16:61–70

MATLAB/Simulink Control Design Toolbox User Manual. MathWorks, Version 4.5 (2017)

Munje R, Patre B (2023) Two-stage multirate state feedback control designs for systems with slow and fast eigenvalue modes. Trans Inst Meas Control 46:1027–1037. https://doi.org/10.1177/01423312231187445

Naidu DS, Price DB, Hibey JL (1987) Singular perturbations and time-scales (SPaTS) in discrete control systems-an overview. In: Proceedings of the IEEE conference on decision and control, vol 26, pp 2096–2103

Naidu DS (1988) Singular perturbation methodology in control systems. Peter Peregrinus Ltd, London U.K

Naidu DS, Rao AK (1985) Singular perturbation analysis of discrete control systems. Springer, Berlin, Heidelberg

Phillips RG (1980) Reduced order modelling and control of two-time-scale discrete systems. Int J Control 31:765–780

Priya PL, Bandyopadhyay B (2017) Discrete-time sliding mode control for uncertain delta operator systems with infrequent output measurements. Eur J Control 33:52–59

Radisavljevic-Gajic V (2015) A simplified two-stage design of linear discrete-time feedback controllers. ASME J Dyn Syst, Meas, Control 137:1–7

Radisavljevic-Gajic V (2015) Two-stage feedback control design for a class of linear discrete-time systems with slow and fast modes. ASME J Dyn Syst, Meas, Control 137:1–6

Satyanarayana N, Janardhanan S (2014) Open-loop functional observer and state feedback for linear systems using multirate output sampling. Asian J Control 16:1897–1902

Tiwari AP, Reddy GD, Bandyopadhyay B (2012) Design of periodic output feedback and fast output sampling based controllers for systems with slow and fast modes. Asian J Control 14:271–277

Zhang Y, Naidu DS, Cai CX, Zou Y (2014) Singular perturbations and time scales in control theories and applications: an overview 2002–2012. Int J Info Syst Sci 9:1–36

# Chapter 6
# Two-Stage Multirate Feedback Control Design Using Observer

**Abstract** This chapter addresses the limitations of single-rate sampling in designing discrete-time state feedback-based controllers for two-time-scale systems. It proposes a multirate state feedback (MRSF) control approach, i.e. sampling slow and fast states at different rates. The method involves constructing a block-triangular form of the continuous two-time-scale system, discretizing with a smaller sampling period for the fast subsystem, and block-diagonalizing for the slow subsystem. A derived MRSF control stabilizes the full-order system, employing a two-stage observer to estimate slow and fast states. Simulation results demonstrate computational savings without compromising closed-loop performance compared to the single-rate sampling method.

**Keywords** Feedback control · Multirate sampling · Singular perturbation · Two-stage design · Two-time-scale system

## 6.1 Introduction

State feedback based multirate controllers designed in the previous chapter require an observer for implementation purposes. In this chapter, an observer-based multirate state feedback (MRSF) control is designed for a two-time-scale system by sampling slow and fast states with different sampling periods in which the slow and fast subsystem states are estimated by a sequential observer (Munje and Zhang 2021). Firstly, the continuous-time system is transformed into an upper triangular form and then feedback controls are designed for two subsystems with different sampling periods. Then, a feedback control input with slow and fast states sampled at different rates is derived. Lastly, for control realization, slow and fast subsystem states are estimated by a sequential two-stage observer (Yoo and Gajic 2018). The presented method reduces design complexity, computations and signal processing time significantly and also improves system performance. The usefulness of the reported design is verified by simulating a numerical example.

 
R. Munje et al., *Multi-stage Feedback Control Design for Multi-time-scale System*,
Springer Tracts in Electrical and Electronics Engineering,
https://doi.org/10.1007/978-981-97-4592-0_6

## 6.2 Design Approach

Consider a linear time-invariant, continuous-time system

$$\dot{\mathbf{z}} = \mathbf{A}\mathbf{z} + \mathbf{B}\mathbf{u}, \tag{6.1}$$

$$\mathbf{y} = \mathbf{C}\mathbf{z}, \tag{6.2}$$

where $\mathbf{z} \in \Re^n$, $\mathbf{u} \in \Re^m$ and $\mathbf{y} \in \Re^p$ are states, inputs and outputs respectively. If the system (6.1)–(6.2) is assumed to have two-time-scale structure, i.e. $n$ eigenvalues have $n_1$ slow and $n_2$ fast eigenvalues then it can be written as

$$\begin{bmatrix} \dot{\mathbf{z}}_1 \\ \dot{\mathbf{z}}_2 \end{bmatrix} = \begin{bmatrix} \mathbf{A}_{11} & \mathbf{A}_{12} \\ \frac{\mathbf{A}_{21}}{\varepsilon} & \frac{\mathbf{A}_{22}}{\varepsilon} \end{bmatrix} \begin{bmatrix} \mathbf{z}_1 \\ \mathbf{z}_2 \end{bmatrix} + \begin{bmatrix} \mathbf{B}_1 \\ \frac{\mathbf{B}_2}{\varepsilon} \end{bmatrix} \mathbf{u}, \tag{6.3}$$

$$\mathbf{y} = \begin{bmatrix} \mathbf{C}_1 & \mathbf{C}_2 \end{bmatrix} \begin{bmatrix} \mathbf{z}_1^T & \mathbf{z}_2^T \end{bmatrix}^T, \tag{6.4}$$

where $\mathbf{z}_1 \in \Re^{n_1}$ and $\mathbf{z}_2 \in \Re^{n_2}$, such that $n_1 + n_2 = n$. Submatrices $\mathbf{A}_{ij}$, $\mathbf{B}_i$ and $\mathbf{C}_i$ are of compatible dimensions. Parameter $\varepsilon$ is the speed ratio of the slow versus fast states.

**Assumption 6.1** The pairs $(\mathbf{A}, \mathbf{B})$ and $(\mathbf{A}, \mathbf{C})$ are respectively controllable and observable (Chen 1999).

**Assumption 6.2** Matrix $\mathbf{A}_{22}$ is invertible (Kokotovic et al. 1999).

Assumption 6.1 guarantees that the state feedback control and the observer for (6.3)–(6.4), respectively, required to satisfy stability and to estimate states. Furthermore, applying similarity transformations, Assumption 6.1 aids in fulfilling the controllability and observability requirements of the lower-order continuous-time subsystems found in the following subsections. For algebraic equation solutions to be discussed further in the design process, Assumption 6.2 is required.

Compared to the fast subsystem states, the slow subsystem states make sense to measure at a slower sampling rate since they are nearly constant throughout a period of $1/\varepsilon$. Slow states can therefore be sampled at the $\tau$ sampling interval if the fast states are sampled at the $\Delta$ sampling interval, where $\Delta = \tau/N$, with $N$ being the largest integer smaller than or equal to $1/\varepsilon$ (Lennarston 1989). Interestingly, for a suitably selected value of a sampling interval, the discrete-time system equivalent to the continuous two-time-scale system (6.3)–(6.4), would also possess two-time-scale property (Naidu and Rao 1985; Munje et al. 2018). Thus, the designer has to select $N$, $\tau$ and $\Delta$. Out of these, $N$ is selected, such that, it is $\leq 1/\varepsilon$. Now, if $\tau$ is selected, one can acquire $\Delta$. Sampling period $\tau$ can be carefully chosen using sampling theorem (Jerri 1977) and the following condition.

**Condition 6.1** If Assumption 6.1 is satisfied and the system (6.3) has only real eigenvalues, then any $T_s > 0$ can be taken, otherwise, if $|\Re e(\psi_i - \psi_j)| = 0$ then

$$T_s \neq \frac{2\pi m}{|\Im m(\psi_i - \psi_j)|}, \qquad \text{for } m = 1, 2, \ldots \tag{6.5}$$

where $\psi_i$ and $\psi_j$ are any two eigenvalues (Chen 1999).

To meet the controllability property of the discrete-time systems obtained by sampling the equivalent continuous-time system, Condition 6.1 must be met. It should be noted that, if $\tau$ satisfies the sampling theorem and Condition 6.1, $\Delta$ also satisfies both.

Suppose the sampling begins at a $k\tau$ instant. Here, slow state variables are measured at every $k\tau$, where $k = 0, 1, 2, \ldots$, while the fast state variables are measured at every $\Delta = k\tau/N$. In other words, in the $\tau$ sampling interval of slow state variables, fast state variables are sampled $N$ times at every $l\Delta$, where $l = 0, 1, 2, \ldots, (N-1)$. The overall feedback control input is the combination of slow and fast controls, such that

$$\mathbf{u}(t) = \mathbf{u}_s(k\tau) + \mathbf{u}_f(k\tau + l\Delta), \tag{6.6}$$

where $\mathbf{u}_s(k\tau)$ and $\mathbf{u}_f(k\tau + l\Delta)$ are piece-wise constant during

$$\begin{cases} \mathbf{u}_s(k\tau) & k\tau \leq t < (k+1)\tau, \\ \mathbf{u}_f(k\tau + l\Delta) & k\tau + l\Delta \leq t < k\tau + (l+1)\Delta, \end{cases}$$

respectively.

In this design, the original continuous-time system is first converted into a block-triangular form to decouple fast subsystem. The block-triangular system is then discretized with a smaller sampling period ($\Delta$). Then, the first stage of feedback control design is carried out. The designed local subsystem controller is applied to the full-order discrete-time system. Later, the slow subsystem is separated and the overall system is equivalently represented into a system with a larger sampling period ($\tau$). In the second stage, feedback control for the slow subsystem is designed. However, slow and fast states of the system (6.3), being internal and decoupled states, cannot be measured and sampled directly for implementing the multirate state feedback control (6.6). For that reason, a sequential observer is designed to estimate them separately. The design of sequential observer is also completed using two-stage design procedure. In the following subsections, the complete design scheme of MRSF control and state estimation by the sequential observer are presented.

### 6.2.1 MRSF Control Design for Two-Time-Scale System

First of all, an upper triangular form of (6.3) is constructed and then its discrete-time forms are acquired sequentially, starting from a lower to higher sampling periods, i.e. $\Delta \rightarrow \tau$. Just after achieving a particular discrete-time form, feedback control for the corresponding uncoupled subsystem is designed. The block-triangular form of (6.3) is given by

$$\begin{bmatrix} \dot{\mathbf{z}}_1 \\ \dot{\mathbf{z}}_f \end{bmatrix} = \begin{bmatrix} \mathbf{A}_s & \mathbf{A}_{12} \\ \mathbf{0} & \frac{\mathbf{A}_f}{\varepsilon} \end{bmatrix} \begin{bmatrix} \mathbf{z}_1 \\ \mathbf{z}_f \end{bmatrix} + \begin{bmatrix} \mathbf{B}_1 \\ \frac{\mathbf{B}_f}{\varepsilon} \end{bmatrix} \mathbf{u}. \tag{6.7}$$

In this, $max|\Re e\{\psi(\mathbf{A}_s)\}| \ll min\left|\Re e\left\{\psi\left(\frac{\mathbf{A}_f}{\varepsilon}\right)\right\}\right|$, where $\psi(.)$ is eigenvalue. The system (6.7) is an outcome of application of

$$\begin{bmatrix} \mathbf{z}_1 \\ \mathbf{z}_f \end{bmatrix} = \begin{bmatrix} \mathbf{I}_{n1} & \mathbf{0} \\ \mathbf{P} & \mathbf{I}_{n2} \end{bmatrix} \begin{bmatrix} \mathbf{z}_1 \\ \mathbf{z}_2 \end{bmatrix}, \tag{6.8}$$

to (6.3), in which $\mathbf{I}_n$ is an identity matrix of order $n$ and matrix $\mathbf{P}$ of order $(n_2 \times n_1)$ satisfies

$$\varepsilon \mathbf{P}(\mathbf{A}_{11} - \mathbf{A}_{12}\mathbf{P}) + \mathbf{A}_{21} - \mathbf{A}_{22}\mathbf{P} = 0, \tag{6.9}$$

where $\mathbf{A}_s = \mathbf{A}_{11} - \mathbf{A}_{12}\mathbf{P}$, $\mathbf{A}_f = \varepsilon\mathbf{P}\mathbf{A}_{12} + \mathbf{A}_{22}$ and $\mathbf{B}_f = \varepsilon\mathbf{P}\mathbf{B}_1 + \mathbf{B}_2$. Solution of (6.9) exists for sufficiently small value of $\varepsilon$ and under Assumption 6.2. Setting $\varepsilon = 0$ in (6.9) gives

$$\mathbf{A}_{21} - \mathbf{A}_{22}\mathbf{P}^{(0)} = 0. \tag{6.10}$$

From (6.10), $\mathbf{P}^{(0)} = \mathbf{A}_{22}^{-1}\mathbf{A}_{21} \Longrightarrow \mathbf{P}^{(0)} = \mathbf{A}_{22}^{-1}\mathbf{A}_{21}$ and $\mathbf{P}^{(0)} = O(1)$ and $\mathbf{P} = \mathbf{P}^{(0)} + O(\varepsilon)$. Having found $\mathbf{P}^{(0)}$, algebraic equation (6.9) can be solved by fixed point iteration method (Naidu 1988). Assumption 6.1 implies system (6.7) is controllable. In (6.7), the fast subsystem is entirely isolated from the slow subsystems. At this point, as per Assumption 6.1, controllability of $\left(\frac{\mathbf{A}_f}{\varepsilon}, \frac{\mathbf{B}_f}{\varepsilon}\right)$ is preserved. Now, the system (6.7) is discretized with $\Delta$ sampling interval as

$$\begin{bmatrix} \mathbf{z}_{1,l+1} \\ \mathbf{z}_{f,l+1} \end{bmatrix} = \begin{bmatrix} \mathbf{\Phi}_{\Delta s} & \mathbf{\Phi}_{\Delta 12} \\ \mathbf{0} & \mathbf{\Phi}_{\Delta f} \end{bmatrix} \begin{bmatrix} \mathbf{z}_{1,l} \\ \mathbf{z}_{f,l} \end{bmatrix} + \begin{bmatrix} \mathbf{\Gamma}_{\Delta 1} \\ \mathbf{\Gamma}_{\Delta f} \end{bmatrix} \mathbf{u}_l, \tag{6.11}$$

where

$$\begin{bmatrix} \mathbf{\Phi}_{\Delta s} & \mathbf{\Phi}_{\Delta 12} \\ \mathbf{0} & \mathbf{\Phi}_{\Delta f} \end{bmatrix} = \bar{\mathbf{\Phi}}_\Delta = e^{\bar{\mathbf{A}}\Delta}, \begin{bmatrix} \Gamma_{\Delta 1} \\ \mathbf{\Gamma}_{\Delta f} \end{bmatrix} = \bar{\Gamma}_\Delta = \int_0^\Delta e^{\bar{\mathbf{A}}\eta}\bar{\mathbf{B}}d\eta,$$

in which

$$\bar{\mathbf{A}} = \begin{bmatrix} \mathbf{A}_s & \mathbf{A}_{12} \\ \mathbf{0} & \frac{\mathbf{A}_f}{\varepsilon} \end{bmatrix} \text{ and } \bar{\mathbf{B}} = \begin{bmatrix} \mathbf{B}_1 \\ \frac{\mathbf{B}_f}{\varepsilon} \end{bmatrix}.$$

Feedback control design for the discrete-time system (6.11) requires following assumption.

**Assumption 6.3** The system $(\bar{\mathbf{\Phi}}_\Delta, \bar{\Gamma}_\Delta)$ is controllable.

If sampling periods ($\tau$ and $\Delta$) meet sampling theorem (Jerri 1977) and Condition 6.1, Assumption 6.3 is convinced. As a result, the controllability of $(\bar{\mathbf{\Phi}}_\Delta, \bar{\Gamma}_\Delta)$ infers the controllability of $(\mathbf{\Phi}_{\Delta f}, \Gamma_{\Delta f})$. And so, in the first stage, feedback control

$$\mathbf{u}_l = -\mathbf{F}_f \mathbf{z}_{f,l} + \mathbf{u}_{s,l}, \tag{6.12}$$

is designed and applied to (6.11) as

$$\begin{bmatrix} \mathbf{z}_{1,l+1} \\ \mathbf{z}_{f,l+1} \end{bmatrix} = \begin{bmatrix} \mathbf{\Phi}_{\Delta s} & \mathbf{\Phi}_{\Delta 12} - \Gamma_{\Delta 1}\mathbf{F}_f \\ \mathbf{0} & \mathbf{\Phi}_{\Delta f} - \Gamma_{\Delta f}\mathbf{F}_f \end{bmatrix} \begin{bmatrix} \mathbf{z}_{1,l} \\ \mathbf{z}_{f,l} \end{bmatrix} + \begin{bmatrix} \mathbf{\Gamma}_{\Delta 1} \\ \mathbf{\Gamma}_{\Delta f} \end{bmatrix} \mathbf{u}_{s,l}. \tag{6.13}$$

The feedback gain $\mathbf{F}_f$ of dimension $(m \times n_2)$ is chosen, such that $\psi(\mathbf{\Phi}_{\Delta f} - \mathbf{\Gamma}_{\Delta f}\mathbf{F}_f) = \psi_{\Delta f}^{desired}$ and positioned near origin. After that, the system (6.13) is modified to

$$\begin{bmatrix} \mathbf{z}_{s,l+1} \\ \mathbf{z}_{f,l+1} \end{bmatrix} = \begin{bmatrix} \mathbf{\Phi}_{\Delta s} & \mathbf{0} \\ \mathbf{0} & \mathbf{\Phi}_{\Delta f} - \mathbf{\Gamma}_{\Delta f}\mathbf{F}_f \end{bmatrix} \begin{bmatrix} \mathbf{z}_{s,l} \\ \mathbf{z}_{f,l} \end{bmatrix} + \begin{bmatrix} \mathbf{\Gamma}_{\Delta s} \\ \mathbf{\Gamma}_{\Delta f} \end{bmatrix} \mathbf{u}_{s,l}, \tag{6.14}$$

to separate fast subsystems. This is achieved by applying

$$\begin{bmatrix} \mathbf{z}_{s,l} \\ \mathbf{z}_{f,l} \end{bmatrix} = \begin{bmatrix} \mathbf{I}_{n1} & -\mathbf{M} \\ \mathbf{0} & \mathbf{I}_{n2} \end{bmatrix} \begin{bmatrix} \mathbf{z}_{1,l} \\ \mathbf{z}_{f,l} \end{bmatrix}, \tag{6.15}$$

to (6.13). In this, the $(n_1 \times n_2)$ matrix $\mathbf{M}$ is evaluated by setting

$$\mathbf{\Phi}_{\Delta 12} - \mathbf{\Gamma}_{\Delta 1}\mathbf{F}_f + \mathbf{\Phi}_{\Delta s}\mathbf{M} - \mathbf{M}(\mathbf{\Phi}_{\Delta f} - \mathbf{\Gamma}_{\Delta f}\mathbf{F}_f) = 0. \tag{6.16}$$

Equation (6.16) can be solved using 'lyap' command of MATLAB (MATLAB/Simulink Control Design Toolbox User Manual 2017) as a linear algebraic Sylvester equation (Chen 1999) with:

**Assumption 6.4** $\psi(\mathbf{\Phi}_{\Delta s}) \neq \psi(\mathbf{\Phi}_{\Delta f} - \mathbf{\Gamma}_{\Delta f}\mathbf{F}_f)$.

Assumption 6.4 is fulfilled because subsystem $(\boldsymbol{\Phi}_{\Delta f} - \boldsymbol{\Gamma}_{\Delta f}\mathbf{F}_f)$ is user designed with the eigenvalues close to the origin, whereas the subsystem $\boldsymbol{\Phi}_{\Delta s}$ has original system's slow eigenvalues. Manipulating (6.13) and (6.15) gives $\boldsymbol{\Gamma}_{\Delta s} = \boldsymbol{\Gamma}_{\Delta 1} - \mathbf{M}\boldsymbol{\Gamma}_{\Delta f}$. The system (6.14) has a sampling period of $\Delta$ seconds. If $\tau = N/\Delta$, system corresponding to $\tau$ sampling interval can be built from (6.14) as

$$\begin{bmatrix} \mathbf{z}_{s,k+1} \\ \mathbf{z}_{f,k+1} \end{bmatrix} = \begin{bmatrix} \boldsymbol{\Phi}_{\tau s} & \mathbf{0} \\ \mathbf{0} & \Phi_{\tau f} \end{bmatrix} \begin{bmatrix} \mathbf{z}_{s,k} \\ \mathbf{z}_{f,k} \end{bmatrix} + \begin{bmatrix} \boldsymbol{\Gamma}_{\tau s} \\ \boldsymbol{\Gamma}_{\tau f} \end{bmatrix} \mathbf{u}_{s,k}. \tag{6.17}$$

The state and input matrices of the $\Delta$ system (6.14) and $\tau$ system (6.17) are related as $\boldsymbol{\Phi}_\tau = \boldsymbol{\Phi}_\Delta^N$ and $\boldsymbol{\Gamma}_\tau = \sum_{i=0}^{N-1} \boldsymbol{\Phi}_\Delta^i \boldsymbol{\Gamma}_\Delta$. As a result, submatrices in (6.17) are obtained as

$$\boldsymbol{\Phi}_{\tau s} = \boldsymbol{\Phi}_{\Delta s}^N, \tag{6.18}$$

$$\boldsymbol{\Phi}_{\tau f} = (\boldsymbol{\Phi}_{\Delta f} - \boldsymbol{\Gamma}_{\Delta f}\mathbf{F}_f)^N, \tag{6.19}$$

$$\boldsymbol{\Gamma}_{\tau s} = \sum_{i=0}^{N-1} \Phi_{\Delta s}^i \boldsymbol{\Gamma}_{\Delta s}, \tag{6.20}$$

$$\boldsymbol{\Gamma}_{\tau f} = \sum_{i=0}^{N-1} (\boldsymbol{\Phi}_{\Delta f} - \boldsymbol{\Gamma}_{\Delta f}\mathbf{F}_f)^i \boldsymbol{\Gamma}_{\Delta f}. \tag{6.21}$$

As the controllability of the slow subsystem $(\boldsymbol{\Phi}_{\tau s}, \boldsymbol{\Gamma}_{\tau s})$ is confirmed by Assumption 6.3, in the second stage

$$\mathbf{u}_{s,k} = -\mathbf{F}_s \mathbf{z}_{s,k}, \tag{6.22}$$

is designed and applied to (6.17) to have a closed-loop system

$$\begin{bmatrix} \mathbf{z}_{s,k+1} \\ \mathbf{z}_{f,k+1} \end{bmatrix} = \begin{bmatrix} \boldsymbol{\Phi}_{\tau s} - \boldsymbol{\Gamma}_{\tau s}\mathbf{F}_s & \mathbf{0} \\ -\boldsymbol{\Gamma}_{\tau f}\mathbf{F}_s & \Phi_{\tau f} \end{bmatrix} \begin{bmatrix} \mathbf{z}_{s,k} \\ \mathbf{z}_{f,k} \end{bmatrix}. \tag{6.23}$$

In this, the feedback gain $\mathbf{F}_s$ of matrix size $(m \times n_1)$ is selected, such that $\psi(\boldsymbol{\Phi}_{\tau s} - \boldsymbol{\Gamma}_{\tau s}\mathbf{F}_s) = \psi_{\tau s}^{desired}$ and are placed near unit circle. Thus, the overall multirate state feedback control is

$$\mathbf{u}(t) = -\mathbf{F}_f \mathbf{z}_{f,l} - \mathbf{F}_s \mathbf{z}_{s,k}, \tag{6.24}$$

where

$$\mathbf{u}_{f,l} = -\mathbf{F}_f \mathbf{z}_{f,l}, \quad k\tau + l\Delta \le t < k\tau + (l+1)\Delta,$$
$$\mathbf{u}_{s,k} = -\mathbf{F}_s \mathbf{z}_{s,k}, \quad k\tau \le t < (k+1)\tau.$$

**Theorem 6.1** *Input (6.24), received by sampling fast and slow subsystem states respectively at $\Delta$ and $\tau$ intervals, stabilizes the system (6.23).*

***Proof*** In a block-triangular system (6.23), eigenvalues are the disjoint sum of eigenvalues of diagonal matrices $(\boldsymbol{\Phi}_{\tau s} - \boldsymbol{\Gamma}_{\tau s}\mathbf{F}_s)$ and $\Phi_{\tau f}$. The subsystem $(\boldsymbol{\Phi}_{\tau s} - \boldsymbol{\Gamma}_{\tau s}\mathbf{F}_s)$ is stabilized with $\mathbf{F}_s$, so that the eigenvalues stay within the unit circle near perimeter. On the other hand, subsystem $\Phi_{\tau f}$ is given by (6.19), in which the eigenvalues of $(\boldsymbol{\Phi}_{\Delta f} - \boldsymbol{\Gamma}_{\Delta f}\mathbf{F}_f)$ are stabilized by the feedback gain $\mathbf{F}_f$. At this instant, if $|\psi(\boldsymbol{\Phi}_{\Delta f} - \boldsymbol{\Gamma}_{\Delta f}\mathbf{F}_f)| < 1$, then

$$|\psi((\boldsymbol{\Phi}_{\Delta f} - \boldsymbol{\Gamma}_{\Delta f}\mathbf{F}_f)^N)| < 1, \tag{6.25}$$

making $|\psi(\Phi_{\tau f})| < 1$. As a deduction, the input (6.24) stabilizes the system (6.23). □

The control input (6.24) is obtained by designing feedback controls for the completely decoupled fast $(\mathbf{z}_f)$ and slow $(\mathbf{z}_s)$ subsystem states at different sampling rates. Being internal and decoupled states, they cannot be measured and sampled directly for implementing (6.24). For that reason, an observer is desired to estimate them separately, so that they can be sampled at different rates. As a result, control (6.24) becomes

$$\mathbf{u}(t) = -\mathbf{F}_f\hat{\mathbf{z}}_{f,l} - \mathbf{F}_s\hat{\mathbf{z}}_{s,k}, \tag{6.26}$$

where $\hat{\mathbf{z}}_f$ and $\hat{\mathbf{z}}_s$ are estimates of the continuous-time fast and slow subsystem states respectively. These are then sampled with $\Delta$ and $\tau$ sampling intervals to get corresponding $\hat{\mathbf{z}}_{f,l}$ and $\hat{\mathbf{z}}_{s,k}$. In the next subsection, estimation of these states is discussed.

### *6.2.2 State Estimation Using Sequential Observer*

The full-order observer for (6.3)–(6.4) can be constructed (Chen 1999) as

$$\begin{aligned}
\begin{bmatrix} \dot{\hat{\mathbf{z}}}_1 \\ \dot{\hat{\mathbf{z}}}_2 \end{bmatrix} &= \begin{bmatrix} \mathbf{A}_{11} & \mathbf{A}_{12} \\ \frac{\mathbf{A}_{21}}{\varepsilon} & \frac{\mathbf{A}_{22}}{\varepsilon} \end{bmatrix} \begin{bmatrix} \hat{\mathbf{z}}_1 \\ \hat{\mathbf{z}}_2 \end{bmatrix} + \begin{bmatrix} \mathbf{B}_1 \\ \frac{\mathbf{B}_2}{\varepsilon} \end{bmatrix} \mathbf{u} + \begin{bmatrix} \mathbf{L}_1 \\ \frac{\mathbf{L}_2}{\varepsilon} \end{bmatrix} (\mathbf{y} - \hat{\mathbf{y}}), \\
\dot{\hat{\mathbf{z}}} &= \mathbf{A}\hat{\mathbf{z}} + \mathbf{B}\mathbf{u} + \mathbf{L}(\mathbf{y} - \hat{\mathbf{y}}) \\
&= (\mathbf{A} - \mathbf{L}\mathbf{C})\hat{\mathbf{z}} + \mathbf{B}\mathbf{u} + \mathbf{L}\mathbf{y}, \qquad (6.27) \\
\hat{\mathbf{y}} &= \begin{bmatrix} \mathbf{C}_1 & \mathbf{C}_2 \end{bmatrix} \begin{bmatrix} \hat{\mathbf{z}}_1^T & \hat{\mathbf{z}}_2^T \end{bmatrix}^T = \mathbf{C}\hat{\mathbf{z}}, \qquad (6.28)
\end{aligned}$$

where $\hat{\mathbf{z}}_1$ and $\hat{\mathbf{z}}_2$ are estimates of $\mathbf{z}_1$ and $\mathbf{z}_2$ respectively. As transpose of the observer feedback matrix $(\mathbf{A}^T - \mathbf{C}^T\mathbf{L}^T)$ generates the dual representation to the system feedback matrix $(\mathbf{A} - \mathbf{BK})$, similar approach (Yoo and Gajic 2018; Munje et al. 2018) can be adapted to design $(\mathbf{A} - \mathbf{BK})$ and $(\mathbf{A} - \mathbf{LC})$. For the same, consider a hypothetical system corresponding to (6.3)–(6.4) as

$$\begin{bmatrix} \dot{\mathbf{x}}_1 \\ \dot{\mathbf{x}}_2 \end{bmatrix} = \begin{bmatrix} \mathbf{A}_{11}^T & \frac{\mathbf{A}_{21}^T}{\varepsilon} \\ \mathbf{A}_{12}^T & \frac{\mathbf{A}_{22}^T}{\varepsilon} \end{bmatrix} \begin{bmatrix} \mathbf{x}_1 \\ \mathbf{x}_2 \end{bmatrix} + \begin{bmatrix} \mathbf{C}_1^T \\ \mathbf{C}_2^T \end{bmatrix} \nu, \tag{6.29}$$

where $\mathbf{x}_1 \in \Re^{n_1}$ and $\mathbf{x}_2 \in \Re^{n_2}$. Introducing the state variable transformation

$$\begin{bmatrix} \mathbf{q}_1 \\ \mathbf{q}_2 \end{bmatrix} = \begin{bmatrix} \mathbf{I}_{n1} & \mathbf{0} \\ \mathbf{0} & \frac{\mathbf{I}_{n2}}{\varepsilon} \end{bmatrix} \begin{bmatrix} \mathbf{x}_1 \\ \mathbf{x}_2 \end{bmatrix} = \mathbf{T}_1^T \begin{bmatrix} \mathbf{x}_1 \\ \mathbf{x}_2 \end{bmatrix}, \tag{6.30}$$

system (6.29) is changed to standard singular perturbation form

$$\begin{bmatrix} \dot{\mathbf{q}}_1 \\ \dot{\mathbf{q}}_2 \end{bmatrix} = \begin{bmatrix} \mathbf{A}_{11}^T & \mathbf{A}_{21}^T \\ \frac{\mathbf{A}_{12}^T}{\varepsilon} & \frac{\mathbf{A}_{22}^T}{\varepsilon} \end{bmatrix} \begin{bmatrix} \mathbf{q}_1 \\ \mathbf{q}_2 \end{bmatrix} + \begin{bmatrix} \mathbf{C}_1^T \\ \frac{\mathbf{C}_2^T}{\varepsilon} \end{bmatrix} \nu. \tag{6.31}$$

Later, the system (6.31) is structured to a lower triangular form

$$\begin{bmatrix} \dot{\mathbf{q}}_s \\ \dot{\mathbf{q}}_2 \end{bmatrix} = \begin{bmatrix} \mathbf{A}_s^T & \mathbf{0} \\ \frac{\mathbf{A}_{12}^T}{\varepsilon} & \frac{\mathbf{A}_f^T}{\varepsilon} \end{bmatrix} \begin{bmatrix} \mathbf{q}_s \\ \mathbf{q}_2 \end{bmatrix} + \begin{bmatrix} \mathbf{C}_s^T \\ \frac{\mathbf{C}_2^T}{\varepsilon} \end{bmatrix} \nu, \tag{6.32}$$

by applying state transformation

$$\begin{bmatrix} \mathbf{q}_s \\ \mathbf{q}_2 \end{bmatrix} = \begin{bmatrix} \mathbf{I}_{n1} & -\varepsilon\mathbf{P}^T \\ \mathbf{0} & \mathbf{I}_{n2} \end{bmatrix} \begin{bmatrix} \mathbf{q}_1 \\ \mathbf{q}_2 \end{bmatrix} = \mathbf{T}_2^T \begin{bmatrix} \mathbf{q}_1 \\ \mathbf{q}_2 \end{bmatrix}, \tag{6.33}$$

to (6.31). Herein, the matrix $\mathbf{P}^T$ of dimension $(n_1 \times n_2)$ is reckoned by solving

$$\varepsilon(\mathbf{A}_{11}^T - \mathbf{P}^T\mathbf{A}_{12}^T)\mathbf{P}^T + \mathbf{A}_{21}^T - \mathbf{P}^T\mathbf{A}_{22}^T = 0, \tag{6.34}$$

iteratively, in the same way as that of (6.9). After that, one can gain $\mathbf{A}_s^T = \mathbf{A}_{11}^T - \mathbf{P}^T\mathbf{A}_{12}^T$, $\mathbf{A}_f^T = \varepsilon\mathbf{A}_{12}^T\mathbf{P}^T + \mathbf{A}_{22}$ and $\mathbf{C}_s^T = \mathbf{C}_1^T - \mathbf{P}^T\mathbf{C}_2^T$. Since, $\psi(\mathbf{A}) = \psi(\mathbf{A}^T)$, the system (6.32) also conforms to the eigenvalue grouping property of the system (6.7). For two-stage observer design, in the first stage, applying

$$\nu = -\mathbf{L}_s^T\mathbf{q}_s + \nu_f, \tag{6.35}$$

to the system (6.32), leads to

$$\begin{bmatrix} \dot{\mathbf{q}}_s \\ \dot{\mathbf{q}}_2 \end{bmatrix} = \begin{bmatrix} \mathbf{A}_s^T - \mathbf{C}_s^T\mathbf{L}_s^T & \mathbf{0} \\ \frac{\mathbf{A}_{12}^T - \mathbf{C}_2^T\mathbf{L}_s^T}{\varepsilon} & \frac{\mathbf{A}_f^T}{\varepsilon} \end{bmatrix} \begin{bmatrix} \mathbf{q}_s \\ \mathbf{q}_2 \end{bmatrix} + \begin{bmatrix} \mathbf{C}_s^T \\ \frac{\mathbf{C}_2^T}{\varepsilon} \end{bmatrix} \nu_f. \tag{6.36}$$

The slow subsystem observer gain $\mathbf{L}_s^T$ of matrix size $(p \times n_1)$ is selected, such that $\psi(\mathbf{A}_s^T - \mathbf{C}_s^T\mathbf{L}_s^T) = \psi_s^{desired}$. As the transformations (6.30) and (6.33) transmutes the original system from (6.29) to (6.32) without losing controllability (or observability), the pair $(\mathbf{A}_s^T, \mathbf{C}_s^T)$ is controllable (or the pair $(\mathbf{A}_s, \mathbf{C}_s)$ is observable), due to which an arbitrary eigenvalue assignment is possible under Assumption 6.1 for slow subsystem. Next, the system (6.36) is restructured as

$$\begin{bmatrix} \dot{\mathbf{q}}_s \\ \dot{\mathbf{q}}_f \end{bmatrix} = \begin{bmatrix} \mathbf{A}_s^T - \mathbf{C}_s^T\mathbf{L}_s^T & \mathbf{0} \\ \mathbf{0} & \frac{\mathbf{A}_f^T}{\varepsilon} \end{bmatrix} \begin{bmatrix} \mathbf{q}_s \\ \mathbf{q}_f \end{bmatrix} + \begin{bmatrix} \mathbf{C}_s^T \\ \frac{\mathbf{C}_f^T}{\varepsilon} \end{bmatrix} \nu_f, \tag{6.37}$$

by applying state variable transformation

$$\begin{bmatrix} \mathbf{q}_s \\ \mathbf{q}_f \end{bmatrix} = \begin{bmatrix} \mathbf{I}_{n1} & \mathbf{0} \\ \mathbf{U}^T & \mathbf{I}_{n2} \end{bmatrix} \begin{bmatrix} \mathbf{q}_s \\ \mathbf{q}_2 \end{bmatrix} = \mathbf{T}_3^T \begin{bmatrix} \mathbf{q}_s \\ \mathbf{q}_2 \end{bmatrix}, \tag{6.38}$$

to the system (6.36), in which $\mathbf{C}_f^T = \mathbf{C}_2^T + \varepsilon\mathbf{H}^T\mathbf{C}_s^T$ and

$$\varepsilon\mathbf{H}^T(\mathbf{A}_s^T - \mathbf{C}_s^T\mathbf{L}_s^T) + \mathbf{A}_{12}^T - \mathbf{C}_2^T\mathbf{L}_s^T - \mathbf{A}_f^T\mathbf{U}^T = 0. \tag{6.39}$$

Just like (6.16), (6.39) can be solved to get $(n_2 \times n_1)$ matrix $\mathbf{H}^T$ with:

**Assumption 6.5** $\psi(\mathbf{A}_s^T - \mathbf{C}_s^T\mathbf{L}_s^T) \neq \psi\left(\frac{\mathbf{A}_f^T}{\varepsilon}\right)$.

Subsystem $(\mathbf{A}_s^T - \mathbf{C}_s^T\mathbf{L}_s^T)$ is user designed with asymptotically stable dominant eigenvalues and $\left(\frac{\mathbf{A}_f^T}{\varepsilon}\right)$ has original system's non-dominant eigenvalues. That being so, Assumption 6.5 is complied. Thereupon, in the second stage, passing input

$$\nu_f = -\mathbf{L}_f^T\mathbf{q}_f, \tag{6.40}$$

to the system (6.37) provides

$$\begin{bmatrix} \dot{\mathbf{q}}_s \\ \dot{\mathbf{q}}_f \end{bmatrix} = \begin{bmatrix} \mathbf{A}_s^T - \mathbf{C}_s^T\mathbf{L}_s^T & -\mathbf{C}_s^T\mathbf{L}_f^T \\ \mathbf{0} & \frac{\mathbf{A}_f^T - \mathbf{C}_f^T\mathbf{L}_f^T}{\varepsilon} \end{bmatrix} \begin{bmatrix} \mathbf{q}_s \\ \mathbf{q}_f \end{bmatrix}. \tag{6.41}$$

The fast subsystem observer gain $\mathbf{L}_f^T$ of dimension $(p \times n_2)$ is designed so as to place $\psi\left(\frac{\mathbf{A}_f^T-\mathbf{C}_f^T\mathbf{L}_f^T}{\varepsilon}\right) = \psi_f^{desired}$. Controllability of the fast subsystem $\left(\frac{\mathbf{A}_f^T}{\varepsilon}, \frac{\mathbf{C}_f^T}{\varepsilon}\right)$ is preserved due to Assumption 6.1. Using inputs (6.35), (6.40) and transformations (6.30), (6.33) and (6.38), the overall observer gain is obtained as

$$\nu = \mathbf{L}^T\mathbf{x} = \left[\mathbf{L}_1^T \ \tfrac{1}{\varepsilon}\mathbf{L}_2^T\right]\left[\mathbf{x}_1^T \ \mathbf{x}_2^T\right]^T, \tag{6.42}$$

where

$$\begin{aligned}\mathbf{L}_1^T &= -\mathbf{L}_s^T - \mathbf{L}_f^T\mathbf{U}^T,\\ \mathbf{L}_2^T &= \varepsilon\mathbf{L}_s^T\mathbf{P}^T + \varepsilon\mathbf{L}_f^T\mathbf{U}^T\mathbf{P}^T - \mathbf{L}_f^T.\end{aligned}$$

Once again, applying change of states

$$\begin{bmatrix}\mathbf{x}_s\\ \mathbf{x}_f\end{bmatrix} = \mathbf{T}_1^{-T}\begin{bmatrix}\mathbf{q}_s\\ \mathbf{q}_f\end{bmatrix}, \tag{6.43}$$

to the system (6.41) results

$$\begin{bmatrix}\dot{\mathbf{x}}_s\\ \dot{\mathbf{x}}_f\end{bmatrix} = \begin{bmatrix}(\mathbf{A}_s - \mathbf{L}_s\mathbf{C}_s)^T & \frac{-(\mathbf{L}_f\mathbf{C}_s)^T}{\varepsilon}\\ \mathbf{0} & \frac{(\mathbf{A}_f-\mathbf{L}_f\mathbf{C}_f)^T}{\varepsilon}\end{bmatrix}\begin{bmatrix}\mathbf{x}_s\\ \mathbf{x}_f\end{bmatrix}. \tag{6.44}$$

The hypothetical system (6.29) and system (6.44) are related by linear transformations (6.30), (6.33), (6.38) and (6.43) as

$$\begin{bmatrix}\mathbf{x}_s\\ \mathbf{x}_f\end{bmatrix} = \mathbf{T}^T\begin{bmatrix}\mathbf{x}_1\\ \mathbf{x}_2\end{bmatrix} \Longrightarrow \begin{bmatrix}\mathbf{x}_1\\ \mathbf{x}_2\end{bmatrix} = \mathbf{T}^{-T}\begin{bmatrix}\mathbf{x}_s\\ \mathbf{x}_f\end{bmatrix}, \tag{6.45}$$

where

$$\mathbf{T}^T = \mathbf{T}_1^{-T}\mathbf{T}_3^T\mathbf{T}_2^T\mathbf{T}_1^T = \begin{bmatrix}\mathbf{I}_{n_1} & -\mathbf{P}^T\\ \varepsilon\mathbf{U}^T & -\varepsilon\mathbf{U}^T\mathbf{P}^T + \mathbf{I}_{n_2}\end{bmatrix}. \tag{6.46}$$

Using (6.44) and the duality principle (Chen 1999; Yoo and Gajic 2018), observer can be configured as

$$\begin{aligned}\begin{bmatrix}\dot{\hat{\mathbf{z}}}_s\\ \dot{\hat{\mathbf{z}}}_f\end{bmatrix} &= \begin{bmatrix}(\mathbf{A}_s - \mathbf{L}_s\mathbf{C}_s) & \mathbf{0}\\ \frac{-(\mathbf{L}_f\mathbf{C}_s)}{\varepsilon} & \frac{(\mathbf{A}_f-\mathbf{L}_f\mathbf{C}_f)}{\varepsilon}\end{bmatrix}\begin{bmatrix}\hat{\mathbf{z}}_s\\ \hat{\mathbf{z}}_f\end{bmatrix}\\ &\quad + \begin{bmatrix}\mathbf{B}_{os}\\ \frac{\mathbf{B}_{of}}{\varepsilon}\end{bmatrix}\mathbf{u} + \begin{bmatrix}\mathbf{L}_s\\ \frac{\mathbf{L}_f}{\varepsilon}\end{bmatrix}\mathbf{y}.\end{aligned} \tag{6.47}$$

Employing (6.45) for two-time-scale system (6.27)–(6.28) gives

$$\begin{bmatrix} \hat{\mathbf{z}}_1 \\ \hat{\mathbf{z}}_2 \end{bmatrix} = \mathbf{T} \begin{bmatrix} \hat{\mathbf{z}}_s \\ \hat{\mathbf{z}}_f \end{bmatrix}, \tag{6.48}$$

resulting into an observer to estimate slow and fast subsystem states as

$$\begin{bmatrix} \dot{\hat{\mathbf{z}}}_s \\ \dot{\hat{\mathbf{z}}}_f \end{bmatrix} = \mathbf{T}^{-1}(\mathbf{A} - \mathbf{LC})\mathbf{T} \begin{bmatrix} \hat{\mathbf{z}}_s \\ \hat{\mathbf{z}}_f \end{bmatrix} + \mathbf{T}^{-1}\mathbf{B}\mathbf{u} + \mathbf{T}^{-1}\mathbf{L}\mathbf{y}. \tag{6.49}$$

Systems (6.47) and (6.49) are equivalent, in which case

$$\mathbf{T}^{-1}\mathbf{L} = \begin{bmatrix} \mathbf{L}_s \\ \frac{\mathbf{L}_f}{\varepsilon} \end{bmatrix} \Longrightarrow \mathbf{L} = \mathbf{T} \begin{bmatrix} \mathbf{L}_s \\ \frac{\mathbf{L}_f}{\varepsilon} \end{bmatrix},$$

can be verified using (6.42) and (6.46). In the similarly manner, using $\mathbf{T}^{-1}\mathbf{B}$, one can get

$$\mathbf{B}_{os} = \mathbf{B}_1 - \varepsilon\mathbf{HPB}_1 - \mathbf{HB}_2, \tag{6.50}$$

$$\mathbf{B}_{of} = \varepsilon\mathbf{PB}_1 + \mathbf{B}_2. \tag{6.51}$$

In the observer (6.47), the slow and fast states are estimated sequentially, in which slow states are independent of the fast states and are used to estimate fast states.

Novelty and benefits of the proposed sequential observer-based MRSF control design are listed below.

1. The MRSF control design is completed using two state variable transformations, compared to three state variable transformations in Phillips (1980). Also, in contrast to four state variable transformations in Yoo and Gajic (2018), only three state variable transformations are required to complete the two-stage design of sequential observer. As a result, manual design complexity is reduced significantly.
2. The offered design requires solutions of two algebraic equations for both MRSF control design and sequential observer design, i.e. solutions of four algebraic equations in all. Perversely, using straightforward designs in Phillips (1980) and Yoo and Gajic (2018), solutions of six algebraic equations are required. As less number of algebraic equations are needed to be solved, the online computations are minimized.
3. The presented control design method uses multirate sampling concept, due to which (i) the online computations are reduced compared to the fast single-rate sampling (Radisavljevic-Gajic 2015), and (ii) the closed-loop performance is greatly improved compared to slower single-rate sampling (Radisavljevic-Gajic 2015).

4. In MRSF control, the online computations are reduced by around $(n_1(N-1))$ in one $\tau$ sampling interval, compared to single-rate sampling control when designed with $\Delta$ seconds.
5. Since, both designs (MRSF control and sequential observer) are done in two independent stages by the application of similarity transformations, eventual inaccuracies made in the second stage will not affect the first stage design accuracy.

## 6.3 Numerical Example

Let us consider a fifth order steam power system (Radisavljevic-Gajic 2015), having two slow and three fast eigenvalues with the state, input and output matrices as

$$\mathbf{A} = \begin{bmatrix} -0.1125 & 0.0468 & 0.1049 & 0.0001 & 0.1876 \\ -0.0001 & -0.1669 & -0.0000 & -0.0002 & 0.6264 \\ -0.0000 & 1.3333 & -2.0001 & -0.0000 & -0.0004 \\ -2.0000 & 0.0001 & -0.0001 & -2.0001 & 0.0001 \\ 0.0006 & -0.0001 & 0.0007 & 1.2677 & -5.0014 \end{bmatrix},$$

$$\mathbf{B} = \begin{bmatrix} 0.0001 \; 0.1499 \; 0.0003 \; 1.5003 \; -0.0022 \end{bmatrix}^T,$$

$$\mathbf{C} = \begin{bmatrix} 1 \; 1 \; 0 \; 0 \; 0 \\ 0 \; 0 \; 1 \; 1 \; 1 \end{bmatrix}.$$

Eigenvalues are located at $(-5.0332, \; -1.9697 \pm j0.1428, \; -0.1542 \pm j0.1494)$ with the ratio of slow versus fast states as

$$\varepsilon = \frac{max|\psi_s|}{min|\psi_f|} = 0.1087.$$

Thus, $N$ is taken as 9 $(< \frac{1}{\varepsilon})$. If the sampling duration for the slow subsystem is selected as $\tau = 1.8$ s, then the fast subsystem states are sampled at every $\Delta = 0.2$ s. Supposing the desired eigenvalues of the continuous-time system be $(-0.20, -0.35, -3.10, -4.30, -5.50)$. Consequently, the desired eigenvalues in $s$-plane for the slow and fast subsystems for $\tau$ and $\Delta$ sampling intervals respectively, can be obtained by $e^{\psi_i T_s}$, where $\psi_i$ is the $i$th eigenvalue of the continuous-time system and $T_s$ is the sampling period. Accordingly, one can get

$$\psi_{\Delta f}^{desired} = (0.5379, 0.4232, 0.3329),$$
$$\psi_{\tau s}^{desired} = (0.6977, 0.5326).$$

Now, solutions of (6.9) is found by fixed point iteration method (Naidu 1988) to get $\mathbf{P}$ as

$$\mathbf{P} = \begin{bmatrix} -0.0685 & -0.7161 \\ 1.0788 & -0.0740 \\ 0.2813 & -0.0268 \end{bmatrix},$$

and an upper triangular form (6.7) is reached with

$$\mathbf{A}_s = \begin{bmatrix} -0.1582 & 0.1269 \\ -0.1761 & -0.1502 \end{bmatrix}, \mathbf{B}_f = \begin{bmatrix} -0.0116 \\ 0.1619 \\ -0.0007 \end{bmatrix}, \mathbf{A}_f = \begin{bmatrix} -0.2183 & 0.0000 & -0.0502 \\ 0.0123 & -0.2175 & 0.0170 \\ 0.0033 & 0.1378 & -0.5399 \end{bmatrix}.$$

This system is then discretized with a sampling period of 0.2 s. Later, to place $\psi_{\Delta f}^{desired}$, the feedback gain is determined as

$$\mathbf{F}_f = \begin{bmatrix} -5.7277 & 1.1959 & 0.8986 \end{bmatrix}.$$

Next, $\mathbf{M}$ is estimated as

$$\mathbf{M} = \begin{bmatrix} -0.1013 & -0.0001 & -0.0231 \\ 0.1094 & -0.1296 & -0.1533 \end{bmatrix},$$

to get (6.14) and subsequently system (6.17) with $\tau$ sampling interval is derived using (6.18)–(6.21), in which

$$\mathbf{\Phi}_{\tau s} = \begin{bmatrix} 0.7250 & 0.1710 \\ -0.2373 & 0.7358 \end{bmatrix}, \mathbf{\Gamma}_{\tau s} = \begin{bmatrix} 0.0458 \\ 0.5401 \end{bmatrix},$$
$$\mathbf{\Phi}_{\tau f} = \begin{bmatrix} 0.0071 & 0.0000 & -0.0011 \\ 0.0348 & 0.0002 & -0.0054 \\ 0.0205 & 0.0004 & -0.0030 \end{bmatrix}, \mathbf{\Gamma}_{\tau f} = \begin{bmatrix} -0.0366 \\ 0.2884 \\ 0.0734 \end{bmatrix}.$$

The slow subsystem eigenvalues are placed at $\psi_{\tau s}^{desired}$ by computing

$$\mathbf{F}_s = \begin{bmatrix} -0.3307 & 0.4548 \end{bmatrix}.$$

Utilizing obtained feedback gains, control input (6.24) is formulated. System initial conditions are taken as $(-0.2, 0.1, 0.1, -0.2, 0.4)$. States $\mathbf{z}_s$ and $\mathbf{z}_f$ are estimated using the sequential observer and are sampled at $\tau$ and $\Delta$ sampling intervals respectively, with desired observer eigenvalues as $(-1.6, -1.8, -9.8, -11.2, -13.1)$. For this, solution of (6.34) is determined and system (6.32) is obtained. $\mathbf{P}^T$ in (6.34) and $\mathbf{A}_s^T$ and $\mathbf{A}_f^T$ in (6.32) are the exact transpose matrices of $\mathbf{P}$ in (6.9) and $\mathbf{A}_s$ and $\mathbf{A}_f$ in (6.7) respectively. Whereas, $\mathbf{C}_s^T$ in (6.32) is obtained as

$$\mathbf{C}_s^T = \begin{bmatrix} 1.0000 & -1.2915 \\ 1.0000 & 0.8168 \end{bmatrix}.$$

After this, to position $\psi_s^{desired}$, the observer gain is determined as

$$\mathbf{L}_s^T = \begin{bmatrix} 0.6363 & 0.9424 \\ -0.6236 & 0.8660 \end{bmatrix}.$$

Then, $\mathbf{U}^T$ is calculated as

$$\mathbf{U}^T = \begin{bmatrix} -20.5110 & 42.0085 \\ -14.6007 & 7.8934 \\ -0.0803 & -5.0421 \end{bmatrix},$$

to gain (6.37) with

$$\mathbf{C}_f^T = \begin{bmatrix} 2.3375 & 7.6117 \\ -0.7293 & 3.7516 \\ -0.5570 & 0.5634 \end{bmatrix}.$$

The fast subsystem eigenvalues are placed at $\psi_f^{desired}$ by determining

$$\mathbf{L}_f^T = \begin{bmatrix} 0.4274 & -1.4710 & 1.4775 \\ -0.0043 & 0.4785 & -0.4947 \end{bmatrix}.$$

Finally, $\mathbf{B}_{os}$ and $\mathbf{B}_{of}$ are calculated using (6.50) and (6.51) respectively as

$$\mathbf{B}_{os} = \begin{bmatrix} 2.1257 \\ -0.6428 \end{bmatrix}, \mathbf{B}_{of} = \begin{bmatrix} -0.0116 \\ 0.1619 \\ -0.0007 \end{bmatrix}.$$

Initial conditions for the estimated states of the original system (6.27) are obtained as given in Yoo and Gajic (2018) and initial conditions for the decoupled states are obtained from the relation (6.48). The entire system is simulated using designed feedback and observer gains, initial conditions, and other submatrices. $\Delta$ and $\tau$ seconds, which represent single-rate sampling, are used to compare the simulation results. Figures 6.1 and 6.2 depict estimation errors of the decoupled and sampled slow and fast subsystem states respectively. The progression of the control input and

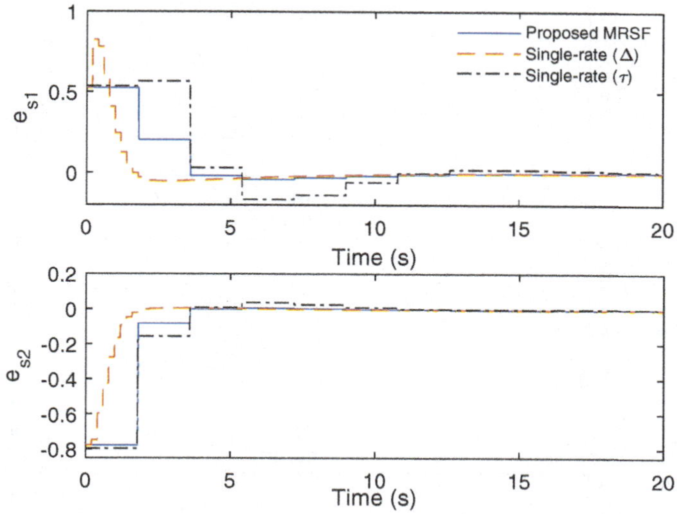


**Fig. 6.1** Estimation errors for the decoupled slow subsystem states

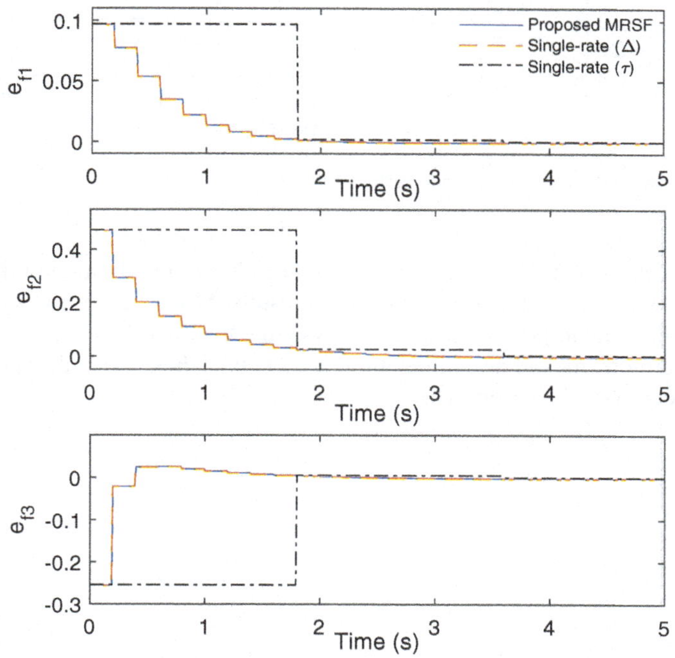


**Fig. 6.2** Estimation errors for the decoupled fast subsystem states

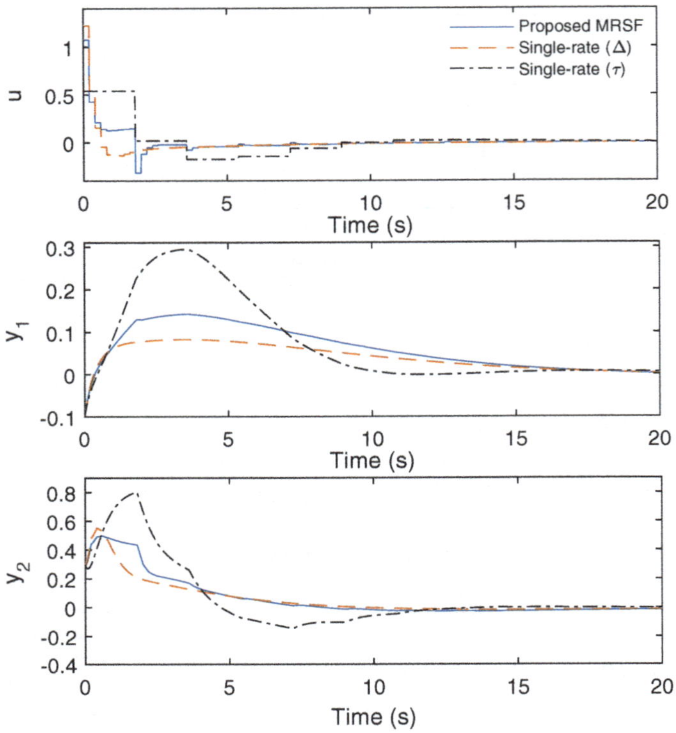


**Fig. 6.3** Evolution of control input and system outputs

outputs is depicted in Fig. 6.3. Figures 6.4 and 6.5 show the evolution of the original system's states. According to the simulation results, MRSF regulation significantly improves the overall performance of the system compared to single-rate sampling ($\tau$). Furthermore, the online computations are less than those of single-rate sampling ($\Delta$) with comparable performance.

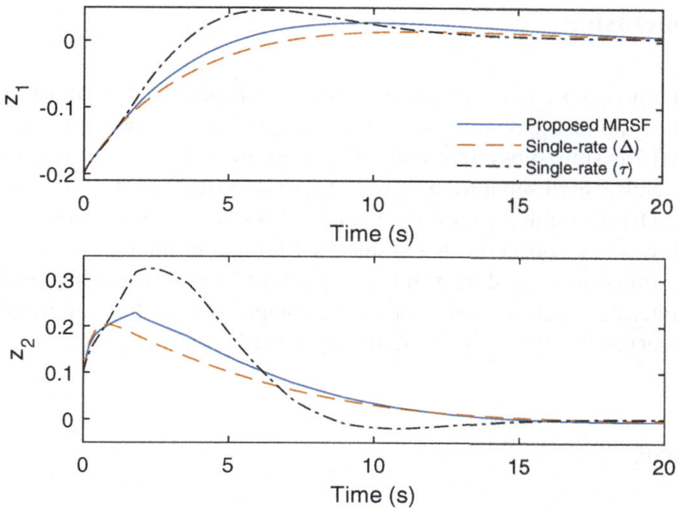


**Fig. 6.4** Evolution of $\mathbf{z}_1$ and $\mathbf{z}_2$ states of the original system

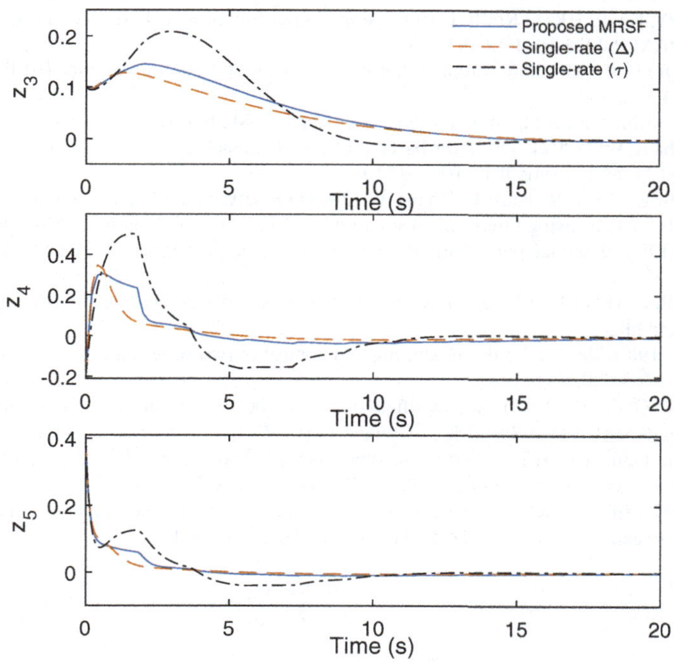


**Fig. 6.5** Evolution of $\mathbf{z}_3$, $\mathbf{z}_4$ and $\mathbf{z}_5$ states of the original system

## 6.4 Conclusion

This chapter presents a multirate state feedback control strategy that uses the precise separation of the fast and slow subsystems to operate on a two-time-scale system. Compared to the slow subsystem states, the rapid subsystem states are sampled more frequently. Subsystem separation in two steps and MRSF control in two stages are combined. The decoupled states are calculated for control implementation using a sequential two-stage observer that is intended for a continuous two-time-scale system. Next, simulations are shown and the suggested control is demonstrated for simulating a numerical example. Simulations are compared with the single-rate sampling and it is observed that the multirate sampling actually improves system performance.

## References

Chen CT (1999) Linear system theory and design. Oxford University Press, New York

Jerri AJ (1977) The Shannon sampling theorem-its various extensions and applications: a tutorial review. Proc IEEE 65:1565–1596

Kokotovic PV, Khalil HK, O'Reilly J (1999) Singular perturbation methods in control: analysis and design. SIAM Publishers, Philadelphia

Lennarston B (1989) Multirate sampled-data control of two-time-scale systems. IEEE Trans Auto Control 34:642–644

MATLAB/Simulink Control Design Toolbox User Manual. MathWorks, Version 4.5 (2017)

Munje R, Zhang WD (2021) Observer-based multirate feedback control design for two-time-scale system. Int J Autom Comput 18:1007–1016

Munje R, Gu D, Desai R, Patre B, Zhang W (2018) Observer-based spatial control of advanced heavy water reactor using time-scale decoupling. IEEE Trans Nucl Sci 65:2756–2766

Naidu DS (1988) Singular perturbation methodology in control systems. Peter Peregrinus Ltd, London U.K

Naidu DS, Rao AK (1985) Singular perturbation analysis of discrete control systems. Springer, Berlin, Heidelberg

Phillips RG (1980) Reduced order modelling and control of two-time-scale discrete systems. Int J Control 31:765–780

Radisavljevic-Gajic V (2015) A simplified two-stage design of linear discrete-time feedback controllers. ASME J Dyn Syst, Meas, Control 137:1–7

Radisavljevic-Gajic V (2015) Two-stage feedback control design for a class of linear discrete-time systems with slow and fast modes. ASME J Dyn Syst, Meas, Control 137:1–6

Yoo H, Gajic Z (2018) New designs of linear observers and observer-based controllers for singularly perturbed linear systems. IEEE Trans Auto Control 63:3904–3911

# Chapter 7
# Three-Stage Multirate Feedback Control Design Using Observer

**Abstract** This chapter addresses the challenges of designing discrete-time state feedback control for a multi-time-scale system, where lower sampling rates fail to capture information from fast and very fast states, while higher rates significantly increase computation. The proposed solution is a multirate state feedback (MRSF) control for a linear time-invariant three-time-scale system, involving sampling of slow, fast, and very fast states at different periods. The method employs a block-triangular form of the continuous three-time-scale system, successive discretization, and simultaneous feedback control design at various sampling rates to formulate a combined MRSF control, proven to stabilize the full-order system. A sequential three-stage observer is introduced for state estimation, and simulations demonstrate the superiority of the approach over single-rate sampling.

**Keywords** Feedback control · Multirate sampling · Sequential observer · Three-stage design · Three-time-scale system

## 7.1 Introduction

In this chapter, an observer-based multirate state feedback (MRSF) control scheme, discussed in Chap. 6, is extended for the three-time-scale system. In this, slow, fast, and very fast states are sampled at different rates and then these are estimated by the sequential three-stage observer for the control realization. Firstly, the continuous-time system is transformed into an upper triangular form and then feedback controls are designed for these three subsystems with different sampling periods. Then, a feedback control input with slow, fast, and very fast states sampled at different rates is derived. Lastly, for control realization, slow, fast, and very fast subsystem states are estimated by a sequential three-stage observer. The presented method minimizes computations and signal processing time significantly and also enhances closed-loop system performance. The applicability of the suggested design is verified by simulating a numerical example.

 
R. Munje et al., *Multi-stage Feedback Control Design for Multi-time-scale System*, Springer Tracts in Electrical and Electronics Engineering,
https://doi.org/10.1007/978-981-97-4592-0_7

## 7.2 Design Strategy

Consider a linear time-invariant, continuous-time system

$$\dot{\mathbf{z}} = \mathbf{A}\mathbf{z} + \mathbf{B}\mathbf{u}, \tag{7.1}$$

$$\mathbf{y} = \mathbf{C}\mathbf{z}, \tag{7.2}$$

where $\mathbf{z} \in \Re^n$, $\mathbf{u} \in \Re^m$ and $\mathbf{y} \in \Re^p$ are states, inputs and outputs respectively. If the system (7.1)–(7.2) is assumed to have three-time-scale structure, i.e., $n$ eigenvalues have $n_1$ slow, $n_2$ fast, and $n_3$ very fast eigenvalues then it can be written as

$$\begin{bmatrix} \dot{\mathbf{z}}_1 \\ \dot{\mathbf{z}}_2 \\ \dot{\mathbf{z}}_3 \end{bmatrix} = \begin{bmatrix} \mathbf{A}_{11} & \mathbf{A}_{12} & \mathbf{A}_{13} \\ \frac{\mathbf{A}_{21}}{\varepsilon} & \frac{\mathbf{A}_{22}}{\varepsilon} & \frac{\mathbf{A}_{23}}{\varepsilon} \\ \frac{\mathbf{A}_{31}}{\mu} & \frac{\mathbf{A}_{32}}{\mu} & \frac{\mathbf{A}_{33}}{\mu} \end{bmatrix} \begin{bmatrix} \mathbf{z}_1 \\ \mathbf{z}_2 \\ \mathbf{z}_3 \end{bmatrix} + \begin{bmatrix} \mathbf{B}_1 \\ \frac{\mathbf{B}_2}{\varepsilon} \\ \frac{\mathbf{B}_3}{\mu} \end{bmatrix} \mathbf{u}, \tag{7.3}$$

$$\mathbf{y} = \begin{bmatrix} \mathbf{C}_1 & \mathbf{C}_2 & \mathbf{C}_3 \end{bmatrix} \begin{bmatrix} \mathbf{z}_1^T & \mathbf{z}_2^T & \mathbf{z}_3^T \end{bmatrix}^T, \tag{7.4}$$

where $\mathbf{z}_1 \in \Re^{n_1}$, $\mathbf{z}_2 \in \Re^{n_2}$ and $\mathbf{z}_3 \in \Re^{n_3}$, such that $n_1 + n_2 + n_3 = n$. Submatrices $\mathbf{A}_{ij}$, $\mathbf{B}_i$ and $\mathbf{C}_i$ are of compatible dimensions. Parameters $\varepsilon$ and $\mu$ are speed ratios of the slow versus fast versus very fast states, such that $\mu \ll \varepsilon \ll 1$.

**Assumption 7.1** The pairs $(\mathbf{A}, \mathbf{B})$ and $(\mathbf{A}, \mathbf{C})$ are respectively controllable and observable (Chen 1999).

**Assumption 7.2** Matrices $\mathbf{A}_{22}$ and $\mathbf{A}_{33}$ are invertible (Kokotovic et al. 1999).

Assumption 7.1 ensures state stabilization and estimation of (7.3)–(7.4) by the state feedback control and observer respectively. It also helps to fulfill the controllability and observability properties of the lower-order continuous-time subsystems obtained in the next subsections by the application of similarity transformations. Assumption 7.2 helps to get solutions of algebraic equations deliberated later in the design procedure.

Let $N_1$ and $N_2$ be integers, such that $N_1 \leq \frac{1}{\varepsilon}$ and $N_2 \leq \frac{\varepsilon}{\mu}$ and $\tau$, $\Delta$ and $\delta$ are sampling periods of slow, fast and very fast states respectively, with $\tau > \Delta > \delta$, then we can introduce

$$\tau = N_1 \Delta = N_1 N_2 \delta. \tag{7.5}$$

The sampling periods $\delta$, $\Delta$, and $\tau$ can be carefully chosen using Conditions 5.1 and 5.2. Suppose the sampling begins at a $k\tau$ instant. Here, slow states are measured at every $k\tau$, where $k = 0, 1, 2, \ldots$, while the fast states are measured at every $k\Delta = k\left(\frac{\tau}{N_1}\right)$ instant and the very fast states are measured at every $k\delta = k\left(\frac{\tau}{N_1 N_2}\right)$ instant. Alternatively, in $\tau$ sampling interval of slow states, fast states are sampled $N_1$ times at every $h\Delta$, where $h = 0, 1, 2, \ldots, (N_1 - 1)$, and in $\Delta$ sampling interval of fast states, very fast states are sampled $N_2$ times at every $l\delta$, where $l = 0, 1, 2, \ldots, (N_2 - 1)$. The overall feedback control is the combination of slow, fast, and very fast controls,

$$\mathbf{u}(t) = \mathbf{u}_s(k\tau) + \mathbf{u}_{f1}(k\tau + h\Delta) + \mathbf{u}_{f2}(k\tau + h\Delta + l\delta), \tag{7.6}$$

where $\mathbf{u}_s(k\tau), \mathbf{u}_{f1}(k\tau + h\Delta)$ and $\mathbf{u}_{f2}(k\tau + h\Delta + l\delta)$ are piece-wise constant during

$$\begin{cases} \mathbf{u}_s(k\tau) & k\tau \leq t < (k+1)\tau, \\ \mathbf{u}_{f1}(k\tau + h\Delta) & k\tau + h\Delta \leq t < k\tau + (h+1)\Delta, \\ \mathbf{u}_{f2}(k\tau + h\Delta + l\delta) & k\tau + h\Delta + l\delta \leq t < k\tau + h\Delta + (l+1)\delta, \end{cases}$$

respectively. In the following subsection, the complete design scheme of MRSF control is proposed.

### 7.2.1 MRSF Control Design for Three-Time-Scale System

First of all, an upper triangular form of (7.3) is constructed and then its discrete-time forms are acquired sequentially, starting from a lower to higher sampling periods, i.e., $\delta \rightarrow \Delta \rightarrow \tau$. Just after achieving a particular discrete-time form, feedback control for the corresponding uncoupled subsystem is designed. The block-triangular form of (7.3) is given by

$$\begin{bmatrix} \dot{\mathbf{z}}_1 \\ \dot{\mathbf{z}}_{fs} \\ \dot{\mathbf{z}}_{f2} \end{bmatrix} = \begin{bmatrix} \mathbf{A}_s & \bar{\mathbf{A}}_{12} & \mathbf{A}_{13} \\ \mathbf{0} & \frac{\mathbf{A}_{f1}}{\varepsilon} & \frac{\bar{\mathbf{A}}_{23}}{\varepsilon} \\ \mathbf{0} & \mathbf{0} & \frac{\mathbf{A}_{f2}}{\mu} \end{bmatrix} \begin{bmatrix} \mathbf{z}_1 \\ \mathbf{z}_{fs} \\ \mathbf{z}_{f2} \end{bmatrix} + \begin{bmatrix} \mathbf{B}_1 \\ \frac{\bar{\mathbf{B}}_{f1}}{\varepsilon} \\ \frac{\mathbf{B}_{f2}}{\mu} \end{bmatrix} \mathbf{u}. \tag{7.7}$$

In this,

$$max|\Re e\{\psi(\mathbf{A}_s)\}| \ll min\left|\Re e\left\{\psi\left(\frac{\mathbf{A}_{f1}}{\varepsilon}\right)\right\}\right|,$$
$$max\left|\Re e\left\{\psi\left(\frac{\mathbf{A}_{f1}}{\varepsilon}\right)\right\}\right| \ll min\left|\Re e\left\{\psi\left(\frac{\mathbf{A}_{f2}}{\mu}\right)\right\}\right|,$$

where $\psi(.)$ is eigenvalue. The system (7.7) is an outcome of application of

$$\begin{bmatrix} \mathbf{z}_1 \\ \mathbf{z}_{fs} \\ \mathbf{z}_{f2} \end{bmatrix} = \begin{bmatrix} \mathbf{I}_{n1} & \mathbf{0} & \mathbf{0} \\ \mathbf{P}_{21} & \mathbf{I}_{n2} & \mathbf{0} \\ \mathbf{P}_{32}\mathbf{P}_{21} + \mathbf{P}_{31} & \mathbf{P}_{32} & \mathbf{I}_{n3} \end{bmatrix} \begin{bmatrix} \mathbf{z}_1 \\ \mathbf{z}_2 \\ \mathbf{z}_3 \end{bmatrix}, \tag{7.8}$$

to (7.3), in which $\mathbf{I}_n$ is an identity matrix of order $n$ and matrices $\mathbf{P}_{21}$ and $\mathbf{P}_{31}$ satisfy following algebraic equations

$$\left.\begin{aligned}\varepsilon \mathbf{P}_{21}\mathbf{A}_{11} + \mathbf{A}_{21} - \mathbf{A}_{f22}\mathbf{P}_{21} - \bar{\mathbf{A}}_{23}\mathbf{P}_{31} &= 0,\\ \mu \mathbf{P}_{31}\mathbf{A}_{11} + \mathbf{A}_{31} - \mathbf{A}_{f32}\mathbf{P}_{21} - \mathbf{A}_{f33}\mathbf{P}_{31} &= 0,\end{aligned}\right\} \tag{7.9}$$

where $\mathbf{A}_s = \mathbf{A}_{11} - \mathbf{A}_{12}\mathbf{P}_{21} - \mathbf{A}_{13}\mathbf{P}_{31}$, $\mathbf{A}_{f22} = \varepsilon\mathbf{P}_{21}\mathbf{A}_{12} + \mathbf{A}_{22}$, $\bar{\mathbf{A}}_{23} = \varepsilon\mathbf{P}_{21}\mathbf{A}_{13} + \mathbf{A}_{23}$, $\mathbf{A}_{f32} = \mu\mathbf{P}_{31}\mathbf{A}_{12} + \mathbf{A}_{32}$, $\mathbf{A}_{f33} = \mu\mathbf{P}_{31}\mathbf{A}_{13} + \mathbf{A}_{33}$, $\bar{\mathbf{B}}_{f1} = \varepsilon\mathbf{P}_{21}\mathbf{B}_1 + \mathbf{B}_2$ and $\bar{\mathbf{B}}_{f2} = \mu\mathbf{P}_{31}\mathbf{B}_1 + \mathbf{B}_3$. Matrix $\mathbf{P}_{32}$ satisfies

$$\frac{\mu}{\varepsilon}\mathbf{P}_{32}\mathbf{A}_{f22} + \mathbf{A}_{f32} - \left(\frac{\mu}{\varepsilon}\mathbf{P}_{32}\bar{\mathbf{A}}_{23} + \mathbf{A}_{f33}\right)\mathbf{P}_{32} = 0. \tag{7.10}$$

with $\bar{\mathbf{A}}_{12} = \mathbf{A}_{12} - \mathbf{A}_{13}\mathbf{P}_{32}$, $\mathbf{A}_{f1} = \mathbf{A}_{f22} - \bar{\mathbf{A}}_{23}\mathbf{P}_{32}$, $\mathbf{A}_{f2} = \frac{\mu}{\varepsilon}\mathbf{P}_{32}\bar{\mathbf{A}}_{23} + \mathbf{A}_{f33}$ and $\mathbf{B}_{f2} = \frac{\mu}{\varepsilon}\mathbf{P}_{32}\bar{\mathbf{B}}_{f1} + \bar{\mathbf{B}}_{f2}$. Solutions of equations in (7.9) exist for sufficiently small values of $\varepsilon$ and under Assumption 7.2. Setting $\varepsilon = 0$ in (7.9) gives

$$\mathbf{A}_{21} - \mathbf{A}_{f22}\mathbf{P}_{21}^{(0)} - \bar{\mathbf{A}}_{23}\mathbf{P}_{31}^{(0)} = 0, \tag{7.11}$$

$$\mathbf{A}_{31} - \mathbf{A}_{f32}\mathbf{P}_{21}^{(0)} - \mathbf{A}_{f33}\mathbf{P}_{31}^{(0)} = 0. \tag{7.12}$$

From (7.11), $\mathbf{P}_{21}^{(0)} = \mathbf{A}_{f22}^{-1}(\mathbf{A}_{21} - \bar{\mathbf{A}}_{23}\mathbf{P}_{31}^{(0)}) \Longrightarrow \mathbf{P}_{21}^{(0)} = \mathbf{A}_{22}^{-1}(\mathbf{A}_{21} - \mathbf{A}_{23}\mathbf{P}_{31}^{(0)})$ and $\mathbf{P}_{21}^{(0)} = O(1)$ and $\mathbf{P}_{21} = \mathbf{P}_{21}^{(0)} + O(\varepsilon)$. Substituting $\mathbf{P}_{21}^{(0)}$ in (7.12) provides $\mathbf{P}_{31}^{(0)} = (\mathbf{A}_{33} - \mathbf{A}_{32}\mathbf{A}_{22}^{-1}\mathbf{A}_{23})^{-1}(\mathbf{A}_{31} - \mathbf{A}_{32}\mathbf{A}_{22}^{-1}\mathbf{A}_{21})$ and $\mathbf{P}_{31}^{(0)} = O(1)$ and $\mathbf{P}_{31} = \mathbf{P}_{31}^{(0)} + O(\varepsilon)$. Since $\mathbf{A}_{f22} = \mathbf{A}_{22} + O(\varepsilon)$ and assuming Assumption 7.2 is satisfied, existence of $\mathbf{A}_{f22}^{-1}$ is ensured. Further, to get $\mathbf{P}_{31}$ following assumption is needed.

**Assumption 7.3** Matrix $(\mathbf{A}_{33} - \mathbf{A}_{32}\mathbf{A}_{22}^{-1}\mathbf{A}_{23})$ is invertible.

Similarly, unique solution of (7.10) for $\mathbf{P}_{32}$ exists for small value of $\mu$ with Assumption 7.2, i.e., for $\mu = 0$, $\mathbf{A}_{f33}\mathbf{P}_{32}^{(0)} = \mathbf{A}_{f32} \Longrightarrow \mathbf{P}_{32}^{(0)} = \mathbf{A}_{33}^{-1}\mathbf{A}_{32} + O(\mu)$. Having found $\mathbf{P}_{21}^{(0)}$, $\mathbf{P}_{31}^{(0)}$ and $\mathbf{P}_{32}^{(0)}$ algebraic equations (7.9) and (7.10) can be solved by fixed point iteration method (Naidu 1988). Assumption 7.1 implies system (7.7) is controllable. In (7.7), the very fast subsystem is entirely isolated from the slow and fast subsystems. At this point, as per Assumption 7.1, controllability of $\left(\frac{\mathbf{A}_{f2}}{\mu}, \frac{\mathbf{B}_{f2}}{\mu}\right)$ is preserved. Now, the system (7.7) is discretized with $\delta$ sampling interval as

$$\begin{bmatrix}\mathbf{z}_{1,l+1}\\ \mathbf{z}_{fs,l+1}\\ \mathbf{z}_{f2,l+1}\end{bmatrix} = \begin{bmatrix}\mathbf{\Phi}_s & \mathbf{\Phi}_{12} & \mathbf{\Phi}_{13}\\ \mathbf{0} & \mathbf{\Phi}_{f1} & \mathbf{\Phi}_{23}\\ \mathbf{0} & \mathbf{0} & \mathbf{\Phi}_{f2}\end{bmatrix}\begin{bmatrix}\mathbf{z}_{1,l}\\ \mathbf{z}_{fs,l}\\ \mathbf{z}_{f2,l}\end{bmatrix} + \begin{bmatrix}\mathbf{\Gamma}_1\\ \bar{\mathbf{\Gamma}}_{f1}\\ \mathbf{\Gamma}_{f2}\end{bmatrix}\mathbf{u}_l, \tag{7.13}$$

where

$$\bar{\mathbf{\Phi}} = \begin{bmatrix}\mathbf{\Phi}_s & \mathbf{\Phi}_{12} & \mathbf{\Phi}_{13}\\ \mathbf{0} & \mathbf{\Phi}_{f1} & \mathbf{\Phi}_{23}\\ \mathbf{0} & \mathbf{0} & \mathbf{\Phi}_{f2}\end{bmatrix} = e^{\bar{\mathbf{A}}\delta},\ \bar{\mathbf{\Gamma}} = \begin{bmatrix}\mathbf{\Gamma}_1\\ \bar{\mathbf{\Gamma}}_{f1}\\ \mathbf{\Gamma}_{f2}\end{bmatrix} = \int_0^{\delta} e^{\bar{\mathbf{A}}\eta}\bar{\mathbf{B}}d\eta,$$

in which

$$\bar{\mathbf{A}} = \begin{bmatrix} \mathbf{A}_s & \bar{\mathbf{A}}_{12} & \mathbf{A}_{13} \\ \mathbf{0} & \frac{\mathbf{A}_{f1}}{\varepsilon} & \frac{\bar{\mathbf{A}}_{23}}{\varepsilon} \\ \mathbf{0} & \mathbf{0} & \frac{\mathbf{A}_{f2}}{\mu} \end{bmatrix} \text{ and } \bar{\mathbf{B}} = \begin{bmatrix} \mathbf{B}_1 \\ \frac{\bar{\mathbf{B}}_{f1}}{\varepsilon} \\ \frac{\mathbf{B}_{f2}}{\mu} \end{bmatrix}.$$

If the original continuous-time system is controllable and has only real eigenvalues, then its discretized version, for $\delta > 0$, is also controllable. However, if the system has complex conjugate eigenvalues then $\delta$ should be appropriately selected as given in Chen (1999). Hence we need the following assumption.

**Assumption 7.4** The system $(\bar{\mathbf{\Phi}}, \bar{\mathbf{\Gamma}})$ is controllable.

Therefore, the subsystem $(\mathbf{\Phi}_{f2}, \mathbf{\Gamma}_{f2})$ is also controllable. And so, in the first stage, feedback control

$$\mathbf{u}_l = -\mathbf{F}_{f2}\mathbf{z}_{f2,l} + \mathbf{u}_{f1,l}, \tag{7.14}$$

is designed and applied to (7.13) as

$$\begin{bmatrix} \mathbf{z}_{1,l+1} \\ \mathbf{z}_{fs,l+1} \\ \mathbf{z}_{f2,l+1} \end{bmatrix} = \begin{bmatrix} \mathbf{\Phi}_s & \mathbf{\Phi}_{12} & \mathbf{\Phi}_{13} - \mathbf{\Gamma}_1\mathbf{F}_{f2} \\ \mathbf{0} & \mathbf{\Phi}_{f1} & \mathbf{\Phi}_{23} - \bar{\mathbf{\Gamma}}_{f1}\mathbf{F}_{f2} \\ \mathbf{0} & \mathbf{0} & \mathbf{\Phi}_{f2} - \mathbf{\Gamma}_{f2}\mathbf{F}_{f2} \end{bmatrix} \begin{bmatrix} \mathbf{z}_{1,l} \\ \mathbf{z}_{fs,l} \\ \mathbf{z}_{f2,l} \end{bmatrix} + \begin{bmatrix} \mathbf{\Gamma}_1 \\ \bar{\mathbf{\Gamma}}_{f1} \\ \mathbf{\Gamma}_{f2} \end{bmatrix} \mathbf{u}_{f1,l}, \tag{7.15}$$

with the purpose of $\psi(\mathbf{\Phi}_{f2} - \mathbf{\Gamma}_{f2}\mathbf{F}_{f2}) = \psi_{\delta f2}^{desired}$. After that, denoting $\bar{\mathbf{\Phi}}_{13} = \mathbf{\Phi}_{13} - \mathbf{\Gamma}_1\mathbf{F}_{f2}$, $\bar{\mathbf{\Phi}}_{23} = \mathbf{\Phi}_{23} - \bar{\mathbf{\Gamma}}_{f1}\mathbf{F}_{f2}$ and $\bar{\mathbf{\Phi}}_{f2} = \mathbf{\Phi}_{f2} - \mathbf{\Gamma}_{f2}\mathbf{F}_{f2}$, system (7.15) is modified to

$$\begin{bmatrix} \mathbf{z}_{1,l+1} \\ \mathbf{z}_{f1,l+1} \\ \mathbf{z}_{f2,l+1} \end{bmatrix} = \begin{bmatrix} \mathbf{\Phi}_s & \mathbf{\Phi}_{12} & \tilde{\mathbf{\Phi}}_{13} \\ \mathbf{0} & \mathbf{\Phi}_{f1} & \mathbf{0} \\ \mathbf{0} & \mathbf{0} & \bar{\mathbf{\Phi}}_{f2} \end{bmatrix} \begin{bmatrix} \mathbf{z}_{1,l} \\ \mathbf{z}_{f1,l} \\ \mathbf{z}_{f2,l} \end{bmatrix} + \begin{bmatrix} \mathbf{\Gamma}_1 \\ \Gamma_{f1} \\ \mathbf{\Gamma}_{f2} \end{bmatrix} \mathbf{u}_{f1,l}, \tag{7.16}$$

to separate fast subsystems. This is achieved by applying

$$\mathbf{z}_{f1,l} = \mathbf{z}_{fs,l} + \mathbf{M}_{23}\mathbf{z}_{f2,l}, \tag{7.17}$$

to (7.15). In this, the matrix $\mathbf{M}_{23}$ is evaluated by setting

$$-\mathbf{\Phi}_{f1}\mathbf{M}_{23} + \bar{\mathbf{\Phi}}_{23} + \mathbf{M}_{23}\bar{\mathbf{\Phi}}_{f2} = 0, \tag{7.18}$$

using 'lyap' command of MATLAB (MATLAB/Simulink Control Design Toolbox User Manual 2017) as a linear algebraic Sylvester equation (Chen 1999) with:

**Assumption 7.5** $\psi(\mathbf{\Phi}_{f1}) \neq \psi(\bar{\mathbf{\Phi}}_{f2})$.

Assumption 7.5 is fulfilled because subsystem $\bar{\mathbf{\Phi}}_{f2}$ is user designed with the eigenvalues close to the origin, whereas the subsystem $\mathbf{\Phi}_{f1}$ has original system's fast eigenvalues. Manipulating (7.15) and (7.17) gives $\tilde{\mathbf{\Phi}}_{13} = -\Phi_{12}\mathbf{M}_{23} + \bar{\mathbf{\Phi}}_{13}$ and

$\Gamma_{f1} = \bar{\mathbf{\Gamma}}_{f1} + \mathbf{M}_{23}\mathbf{\Gamma}_{f2}$. The system (7.16) has a sampling period of $\delta$ seconds. If $\Delta = N_2\delta$, system corresponding to $\Delta$ sampling interval can be built from (7.16) as

$$\begin{bmatrix} \mathbf{z}_{1,h+1} \\ \mathbf{z}_{f1,h+1} \\ \mathbf{z}_{f2,h+1} \end{bmatrix} = \begin{bmatrix} \mathbf{\Phi}_{\Delta s} & \mathbf{\Phi}_{\Delta 12} & \Phi_{\Delta 13} \\ \mathbf{0} & \mathbf{\Phi}_{\Delta f1} & \mathbf{0} \\ \mathbf{0} & \mathbf{0} & \Phi_{\Delta f2} \end{bmatrix} \begin{bmatrix} \mathbf{z}_{1,h} \\ \mathbf{z}_{f1,h} \\ \mathbf{z}_{f2,h} \end{bmatrix} + \begin{bmatrix} \mathbf{\Gamma}_{\Delta 1} \\ \mathbf{\Gamma}_{\Delta f1} \\ \mathbf{\Gamma}_{\Delta f2} \end{bmatrix} \mathbf{u}_{f1,h}. \tag{7.19}$$

The system and input matrices of $\delta$ system (7.16) and $\Delta$ system (7.19) are related as $\mathbf{\Phi}_\Delta = \mathbf{\Phi}_\delta^{N_2}$ and $\mathbf{\Gamma}_\Delta = \sum_{i=0}^{N_2-1} \mathbf{\Phi}_\delta^i \mathbf{\Gamma}_\delta$. Using this relation, submatrices in (7.19) are calculated as

$$\left.\begin{aligned} \mathbf{\Phi}_{\Delta s} &= \mathbf{\Phi}_s^{N_2}, \mathbf{\Phi}_{\Delta f1} = \mathbf{\Phi}_{f1}^{N_2}, \mathbf{\Phi}_{\Delta f2} = \bar{\mathbf{\Phi}}_{f2}^{N_2}, \\ \Phi_{\Delta 12} &= \sum_{i=0}^{N_2-1} \left( \mathbf{\Phi}_s^{N_2-1-i} \Phi_{12} \Phi_{f1}^i \right), \\ \Phi_{\Delta 13} &= \sum_{i=0}^{N_2-1} \left( \mathbf{\Phi}_s^{N_2-1-i} \tilde{\mathbf{\Phi}}_{13} \bar{\mathbf{\Phi}}_{f2}^i \right), \\ \mathbf{\Gamma}_{\Delta 1} &= \sum_{i=0}^{N_2-1} \mathbf{\Phi}_s^i \mathbf{\Gamma}_1 + \sum_{k=1}^{N_2-1} \left\{ \sum_{i=0}^{k-1} \mathbf{\Phi}_s^{k-1-i} \right. \\ &\quad \left. \left( \Phi_{12} \Phi_{f1}^i \Gamma_{f1} + \tilde{\mathbf{\Phi}}_{13} \bar{\mathbf{\Phi}}_{f2}^i \mathbf{\Gamma}_{f2} \right) \right\}, \\ \mathbf{\Gamma}_{\Delta f1} &= \sum_{i=0}^{N_2-1} \mathbf{\Phi}_{f1}^i \Gamma_{f1}, \mathbf{\Gamma}_{\Delta f2} = \sum_{i=0}^{N_2-1} \bar{\mathbf{\Phi}}_{f2}^i \mathbf{\Gamma}_{f2}. \end{aligned} \right\} \tag{7.20}$$

As an appropriate change in the sampling period does not change the linear system properties, the controllability of the updated fast subsystem $(\mathbf{\Phi}_{\Delta f1}, \mathbf{\Gamma}_{\Delta f1})$ is retained. An input $\mathbf{u}_{f1,l}$ in (7.14), at $\delta$ sampling interval, is now changed to the $\mathbf{u}_{f1,h}$ at $\Delta$ sampling interval in (7.19). In the second stage, the control input for the fast subsystem is designed as

$$\mathbf{u}_{f1,h} = -\mathbf{F}_{f1}\mathbf{z}_{f1,h} + \mathbf{u}_{s,h}, \tag{7.21}$$

where $\mathbf{u}_{s,h}$ is a slow subsystem input at $\Delta$ sampling period, which later will be represented at $\tau$ sampling interval. The input (7.21) is passed to (7.19) to have a closed-loop system as

$$\begin{bmatrix} \mathbf{z}_{1,h+1} \\ \mathbf{z}_{f1,h+1} \\ \mathbf{z}_{f2,h+1} \end{bmatrix} = \begin{bmatrix} \mathbf{\Phi}_{\Delta s} & \mathbf{\Phi}_{\Delta 12} - \mathbf{\Gamma}_{\Delta 1}\mathbf{F}_{f1} & \Phi_{\Delta 13} \\ \mathbf{0} & \mathbf{\Phi}_{\Delta f1} - \mathbf{\Gamma}_{\Delta f1}\mathbf{F}_{f1} & \mathbf{0} \\ \mathbf{0} & -\mathbf{\Gamma}_{\Delta f2}\mathbf{F}_{f1} & \Phi_{\Delta f2} \end{bmatrix} \begin{bmatrix} \mathbf{z}_{1,h} \\ \mathbf{z}_{f1,h} \\ \mathbf{z}_{f2,h} \end{bmatrix} + \begin{bmatrix} \mathbf{\Gamma}_{\Delta 1} \\ \mathbf{\Gamma}_{\Delta f1} \\ \mathbf{\Gamma}_{\Delta f2} \end{bmatrix} \mathbf{u}_{s,h}, \tag{7.22}$$

so that $\psi(\mathbf{\Phi}_{\Delta f1} - \mathbf{\Gamma}_{\Delta f1}\mathbf{F}_{f1}) = \psi_{\Delta f1}^{desired}$. Defining $\bar{\mathbf{\Phi}}_{\Delta 12} = \mathbf{\Phi}_{\Delta 12} - \mathbf{\Gamma}_{\Delta 1}\mathbf{F}_{f1}$, $\bar{\mathbf{\Phi}}_{\Delta f1} = \mathbf{\Phi}_{\Delta f1} - \mathbf{\Gamma}_{\Delta f1}\mathbf{F}_{f1}$ and $\bar{\mathbf{\Phi}}_{\Delta 32} = -\mathbf{\Gamma}_{\Delta f2}\mathbf{F}_{f1}$, the system (7.22) is transformed into

$$\begin{bmatrix} \mathbf{z}_{s,h+1} \\ \mathbf{z}_{f1,h+1} \\ \mathbf{z}_{f2,h+1} \end{bmatrix} = \begin{bmatrix} \mathbf{\Phi}_{\Delta s} & \mathbf{0} & \mathbf{0} \\ \mathbf{0} & \bar{\mathbf{\Phi}}_{\Delta f1} & \mathbf{0} \\ \mathbf{0} & \bar{\mathbf{\Phi}}_{\Delta 32} & \Phi_{\Delta f2} \end{bmatrix} \begin{bmatrix} \mathbf{z}_{s,h} \\ \mathbf{z}_{f1,h} \\ \mathbf{z}_{f2,h} \end{bmatrix} + \begin{bmatrix} \mathbf{\Gamma}_{\Delta s} \\ \mathbf{\Gamma}_{\Delta f1} \\ \mathbf{\Gamma}_{\Delta f2} \end{bmatrix} \mathbf{u}_{s,h}. \tag{7.23}$$

to decouple slow subsystem. This is done by applying

$$\mathbf{z}_{s,h} = \mathbf{z}_{1,h} + \mathbf{M}_{12}\mathbf{z}_{f1,h} + \mathbf{M}_{13}\mathbf{z}_{f2,h}, \tag{7.24}$$

to (7.22). Matrices $\mathbf{M}_{12}$ and $\mathbf{M}_{13}$ are found from

$$-\boldsymbol{\Phi}_{\Delta s}\mathbf{M}_{13} + \boldsymbol{\Phi}_{\Delta 13} + \mathbf{M}_{13}\Phi_{\Delta f2} = 0, \tag{7.25}$$

$$-\boldsymbol{\Phi}_{\Delta s}\mathbf{M}_{12} + \bar{\boldsymbol{\Phi}}_{\Delta 12} + \mathbf{M}_{12}\bar{\boldsymbol{\Phi}}_{\Delta f1} + \mathbf{M}_{13}\bar{\boldsymbol{\Phi}}_{\Delta 32} = 0. \tag{7.26}$$

Equations (7.25) and (7.26) can be solved independently, similar to (7.18). First, (7.25) is solved to get $\mathbf{M}_{13}$, which requires

**Assumption 7.6** $\psi(\boldsymbol{\Phi}_{\Delta s}) \neq \psi(\Phi_{\Delta f2})$.

Having obtained $\mathbf{M}_{13}$, (7.26) is solved to get $\mathbf{M}_{12}$, which requires the next assumption

**Assumption 7.7** $\psi(\boldsymbol{\Phi}_{\Delta s}) \neq \psi(\bar{\boldsymbol{\Phi}}_{\Delta f1})$.

Assumptions 7.6 and 7.7 are convinced by the fact that the closed-loop subsystems $\Phi_{\Delta f2}$ and $\bar{\boldsymbol{\Phi}}_{\Delta f1}$ are designed with the stable very fast and fast eigenvalues respectively, while $\boldsymbol{\Phi}_{\Delta s}$ has original system's slow eigenvalues. Application of (7.24) to (7.22) gives $\boldsymbol{\Gamma}_{\Delta s} = \boldsymbol{\Gamma}_{\Delta 1} + \mathbf{M}_{12}\boldsymbol{\Gamma}_{\Delta f1} + \mathbf{M}_{13}\boldsymbol{\Gamma}_{\Delta f2}$. Changing the states of (7.22) to (7.23) using (7.24) alters it from one form to another holding the controllability of the uncoupled slow subsystem $(\boldsymbol{\Phi}_{\Delta s}, \boldsymbol{\Gamma}_{\Delta s})$ as it is, by Assumption 7.4. The system (7.23) has a sampling period of $\Delta$ s. Using (7.5), an equivalent system at $\tau$ sampling interval can be developed from (7.23) as

$$\begin{bmatrix} \mathbf{z}_{s,k+1} \\ \mathbf{z}_{f1,k+1} \\ \mathbf{z}_{f2,k+1} \end{bmatrix} = \begin{bmatrix} \boldsymbol{\Phi}_{\tau s} & \mathbf{0} & \mathbf{0} \\ \mathbf{0} & \Phi_{\tau f1} & \mathbf{0} \\ \mathbf{0} & \Phi_{\tau 32} & \Phi_{\tau f2} \end{bmatrix} \begin{bmatrix} \mathbf{z}_{s,k} \\ \mathbf{z}_{f1,k} \\ \mathbf{z}_{f2,k} \end{bmatrix} + \begin{bmatrix} \boldsymbol{\Gamma}_{\tau s} \\ \boldsymbol{\Gamma}_{\tau f1} \\ \boldsymbol{\Gamma}_{\tau f2} \end{bmatrix} \mathbf{u}_{s,k}. \tag{7.27}$$

Here, system and input matrices of the updated $\Delta$ system (7.23) and $\tau$ system (7.27) are related as $\boldsymbol{\Phi}_\tau = \boldsymbol{\Phi}_{\Delta_u}^{N_1}$ and $\boldsymbol{\Gamma}_\tau = \sum_{i=0}^{N_1-1} \boldsymbol{\Phi}_{\Delta_u}^i \boldsymbol{\Gamma}_{\Delta_u}$. As a result, submatrices in (7.27) are obtained as

$$\left.\begin{aligned} \boldsymbol{\Phi}_{\tau s} &= \boldsymbol{\Phi}_{\Delta s}^{N_1}, \boldsymbol{\Phi}_{\tau f1} = \bar{\boldsymbol{\Phi}}_{\Delta f1}^{N_1}, \boldsymbol{\Phi}_{\tau f2} = \Phi_{\Delta f2}^{N_1}, \\ \Phi_{\tau 32} &= \sum_{i=0}^{N_1-1} (\boldsymbol{\Phi}_{\Delta f2}^{N_1-1-i} \bar{\boldsymbol{\Phi}}_{\Delta 32} \bar{\boldsymbol{\Phi}}_{\Delta f1}^{i}), \\ \boldsymbol{\Gamma}_{\tau s} &= \sum_{i=0}^{N_1-1} \Phi_{\Delta s}^{i} \boldsymbol{\Gamma}_{\Delta s}, \\ \boldsymbol{\Gamma}_{\tau f1} &= \sum_{i=0}^{N_1-1} \bar{\boldsymbol{\Phi}}_{\Delta f1}^{i} \boldsymbol{\Gamma}_{\Delta f1}, \\ \boldsymbol{\Gamma}_{\tau f2} &= \sum_{i=0}^{N_1-1} \boldsymbol{\Phi}_{\Delta f2}^{i} \boldsymbol{\Gamma}_{\Delta f2} + \sum_{k=1}^{N_1-1} \\ &\quad \left\{ \sum_{i=0}^{k-1} \left( \boldsymbol{\Phi}_{\Delta f2}^{k-1-i} \bar{\boldsymbol{\Phi}}_{\Delta 32} \bar{\boldsymbol{\Phi}}_{\Delta f1}^{i} \right) \boldsymbol{\Gamma}_{\Delta f1} \right\}. \end{aligned}\right\} \tag{7.28}$$

Assuming that the change in sampling rate does not change the controllability of the slow subsystem $(\mathbf{\Phi}_{\tau s}, \mathbf{\Gamma}_{\tau s})$, in the third stage

$$\mathbf{u}_{s,k} = -\mathbf{F}_s \mathbf{z}_{s,k}, \tag{7.29}$$

is designed and applied to (7.27) to have a closed-loop system

$$\begin{bmatrix} \mathbf{z}_{s,k+1} \\ \mathbf{z}_{f1,k+1} \\ \mathbf{z}_{f2,k+1} \end{bmatrix} = \begin{bmatrix} \mathbf{\Phi}_{\tau s} - \mathbf{\Gamma}_{\tau s}\mathbf{F}_s & \mathbf{0} & \mathbf{0} \\ -\mathbf{\Gamma}_{\tau f1}\mathbf{F}_s & \Phi_{\tau f1} & \mathbf{0} \\ -\mathbf{\Gamma}_{\tau f2}\mathbf{F}_s & \Phi_{\tau 32} & \Phi_{\tau f2} \end{bmatrix} \begin{bmatrix} \mathbf{z}_{s,k} \\ \mathbf{z}_{f1,k} \\ \mathbf{z}_{f2,k} \end{bmatrix}, \tag{7.30}$$

so as to locate $\psi(\mathbf{\Phi}_{\tau s} - \mathbf{\Gamma}_{\tau s}\mathbf{F}_s) = \psi_{\tau s}^{desired}$. Thus, the overall multirate state feedback control is

$$\mathbf{u}(t) = -\mathbf{F}_{f2}\mathbf{z}_{f2,l} - \mathbf{F}_{f1}\mathbf{z}_{f1,h} - \mathbf{F}_s\mathbf{z}_{s,k}, \tag{7.31}$$

where

$$\begin{aligned} \mathbf{u}_{f2,l} &= -\mathbf{F}_{f2}\mathbf{z}_{f2,l}, \quad k\tau + h\Delta + l\delta \le t < k\tau + h\Delta + (l+1)\delta, \\ \mathbf{u}_{f1,h} &= -\mathbf{F}_{f1}\mathbf{z}_{f1,h}, \qquad k\tau + h\Delta \le t < k\tau + (h+1)\Delta, \\ \mathbf{u}_{s,k} &= -\mathbf{F}_s\mathbf{z}_{s,k}, \qquad\qquad k\tau \le t < (k+1)\tau. \end{aligned}$$

**Theorem 7.1** *Input (7.31), received by sampling very fast, fast and slow subsystem states respectively at $\delta$, $\Delta$ and $\tau$ intervals, stabilizes (7.30).*

***Proof*** In a block-triangular system (7.30), eigenvalues are the disjoint sum of eigenvalues of diagonal matrices $(\mathbf{\Phi}_{\tau s} - \mathbf{\Gamma}_{\tau s}\mathbf{F}_s)$, $\Phi_{\tau f1}$ and $\Phi_{\tau f2}$. The subsystem $(\mathbf{\Phi}_{\tau s} - \mathbf{\Gamma}_{\tau s}\mathbf{F}_s)$ is stabilized with $\mathbf{F}_s$ so that the eigenvalues stay within the unit circle near the perimeter. On the other hand, subsystem $\Phi_{\tau f1}$ is given by (7.28), in which the eigenvalues of $(\mathbf{\Phi}_{\Delta f1} - \mathbf{\Gamma}_{\Delta f1}\mathbf{F}_{f1})$ are stabilized by the feedback gain $\mathbf{F}_{f1}$. At this instant, if $|\psi(\mathbf{\Phi}_{\Delta f1} - \mathbf{\Gamma}_{\Delta f1}\mathbf{F}_{f1})| < 1$, then

$$|\psi((\mathbf{\Phi}_{\Delta f1} - \mathbf{\Gamma}_{\Delta f1}\mathbf{F}_{f1})^{N_1})| < 1, \tag{7.32}$$

making $|\psi(\Phi_{\tau f1})| < 1$. Likewise, the subsystem $\Phi_{\tau f2}$ is given by (7.20) and (7.28), in which

$$|\psi(\mathbf{\Phi}_{f2} - \mathbf{\Gamma}_{f2}\mathbf{F}_{f2})| < 1. \tag{7.33}$$

If (7.33) is satisfied then

$$|\psi((\mathbf{\Phi}_{f2} - \mathbf{\Gamma}_{f2}\mathbf{F}_{f2})^{N_2 N_1})| < 1, \text{ i.e., } |\psi(\Phi_{\tau f2})| < 1. \tag{7.34}$$

As a deduction, the input (7.31) stabilizes the system (7.30). □

The control input (7.31) is obtained by designing feedback controls for the completely decoupled very fast ($\mathbf{z}_{f2}$), fast ($\mathbf{z}_{fs}$) and slow ($\mathbf{z}_s$) states at different sampling rates. Being internal and decoupled states, they cannot be measured and sampled directly for implementing (7.31). For that reason, an observer is desired to estimate them separately, so that they can be sampled at different rates. As a result, control (7.31) has a form

$$\mathbf{u}(t) = -\mathbf{F}_{f2}\hat{\mathbf{z}}_{f2,l} - \mathbf{F}_{f1}\hat{\mathbf{z}}_{f1,h} - \mathbf{F}_s\hat{\mathbf{z}}_{s,k}, \tag{7.35}$$

where $\hat{\mathbf{z}}_{f2}$, $\hat{\mathbf{z}}_{f1}$, and $\hat{\mathbf{z}}_s$ are the estimated continuous-time states, which are sampled at $\delta$, $\Delta$ and $\tau$ intervals respectively.

### 7.2.2 *Three-Stage Design of Sequential Observer*

Full-order observer for (7.3)–(7.4) can be constructed (Chen 1999) as

$$\begin{bmatrix} \dot{\hat{\mathbf{z}}}_1 \\ \dot{\hat{\mathbf{z}}}_2 \\ \dot{\hat{\mathbf{z}}}_3 \end{bmatrix} = \begin{bmatrix} \mathbf{A}_{11} & \mathbf{A}_{12} & \mathbf{A}_{13} \\ \frac{\mathbf{A}_{21}}{\varepsilon} & \frac{\mathbf{A}_{22}}{\varepsilon} & \frac{\mathbf{A}_{23}}{\varepsilon} \\ \frac{\mathbf{A}_{31}}{\mu} & \frac{\mathbf{A}_{32}}{\mu} & \frac{\mathbf{A}_{33}}{\mu} \end{bmatrix} \begin{bmatrix} \hat{\mathbf{z}}_1 \\ \hat{\mathbf{z}}_2 \\ \hat{\mathbf{z}}_3 \end{bmatrix} + \begin{bmatrix} \mathbf{B}_1 \\ \frac{\mathbf{B}_2}{\varepsilon} \\ \frac{\mathbf{B}_3}{\mu} \end{bmatrix} \mathbf{u} + \begin{bmatrix} \mathbf{L}_1 \\ \frac{\mathbf{L}_2}{\varepsilon} \\ \frac{\mathbf{L}_3}{\mu} \end{bmatrix} (\mathbf{y} - \hat{\mathbf{y}}),$$

$$= \mathbf{A}\hat{\mathbf{z}} + \mathbf{B}\mathbf{u} + \mathbf{L}(\mathbf{y} - \hat{\mathbf{y}}) = (\mathbf{A} - \mathbf{L}\mathbf{C})\hat{\mathbf{z}} + \mathbf{B}\mathbf{u} + \mathbf{L}\mathbf{y}, \tag{7.36}$$

$$\hat{\mathbf{y}} = \begin{bmatrix} \mathbf{C}_1 & \mathbf{C}_2 & \mathbf{C}_3 \end{bmatrix} \begin{bmatrix} \hat{\mathbf{z}}_1^T & \hat{\mathbf{z}}_2^T & \hat{\mathbf{z}}_3^T \end{bmatrix}^T = \mathbf{C}\hat{\mathbf{z}}, \tag{7.37}$$

where $\hat{\mathbf{z}}_1$, $\hat{\mathbf{z}}_2$, and $\hat{\mathbf{z}}_3$ are estimates of $\mathbf{z}_1$, $\mathbf{z}_2$, and $\mathbf{z}_3$ respectively. As transpose of the observer feedback matrix ($\mathbf{A}^T - \mathbf{C}^T\mathbf{L}^T$) generates the dual representation to the system feedback matrix ($\mathbf{A} - \mathbf{BK}$), similar approach (Radisavljevic-Gajic et al. 2017; Yoo and Gajic 2018; Munje et al. 2018) can be adapted to design ($\mathbf{A} - \mathbf{BK}$) and ($\mathbf{A} - \mathbf{LC}$). For the same, consider a hypothetical system corresponding to (7.3)–(7.4) as

$$\begin{bmatrix} \dot{\mathbf{x}}_1 \\ \dot{\mathbf{x}}_2 \\ \dot{\mathbf{x}}_3 \end{bmatrix} = \begin{bmatrix} \mathbf{A}_{11}^T & \frac{\mathbf{A}_{21}^T}{\varepsilon} & \frac{\mathbf{A}_{31}^T}{\mu} \\ \mathbf{A}_{12}^T & \frac{\mathbf{A}_{22}^T}{\varepsilon} & \frac{\mathbf{A}_{32}^T}{\mu} \\ \mathbf{A}_{13}^T & \frac{\mathbf{A}_{23}^T}{\varepsilon} & \frac{\mathbf{A}_{33}^T}{\mu} \end{bmatrix} \begin{bmatrix} \mathbf{x}_1 \\ \mathbf{x}_2 \\ \mathbf{x}_3 \end{bmatrix} + \begin{bmatrix} \mathbf{C}_1^T \\ \mathbf{C}_2^T \\ \mathbf{C}_3^T \end{bmatrix} \nu, \tag{7.38}$$

where $\mathbf{x}_1 \in \Re^{n_1}$, $\mathbf{x}_2 \in \Re^{n_2}$ and $\mathbf{x}_3 \in \Re^{n_3}$. Introducing

$$\begin{bmatrix} \mathbf{q}_1 \\ \mathbf{q}_2 \\ \mathbf{q}_3 \end{bmatrix} = \begin{bmatrix} \mathbf{I}_{n1} & \mathbf{0} & \mathbf{0} \\ \mathbf{0} & \frac{\mathbf{I}_{n2}}{\varepsilon} & \mathbf{0} \\ \mathbf{0} & \mathbf{0} & \frac{\mathbf{I}_{n3}}{\mu} \end{bmatrix} \begin{bmatrix} \mathbf{x}_1 \\ \mathbf{x}_2 \\ \mathbf{x}_3 \end{bmatrix} = \mathbf{T}_1^T \begin{bmatrix} \mathbf{x}_1 \\ \mathbf{x}_2 \\ \mathbf{x}_3 \end{bmatrix}, \tag{7.39}$$

system (7.38) is changed to standard singular perturbed form

$$\begin{bmatrix} \dot{\mathbf{q}}_1 \\ \dot{\mathbf{q}}_2 \\ \dot{\mathbf{q}}_3 \end{bmatrix} = \begin{bmatrix} \mathbf{A}_{11}^T & \mathbf{A}_{21}^T & \mathbf{A}_{31}^T \\ \frac{\mathbf{A}_{12}^T}{\varepsilon} & \frac{\mathbf{A}_{22}^T}{\varepsilon} & \frac{\mathbf{A}_{32}^T}{\varepsilon} \\ \frac{\mathbf{A}_{13}^T}{\mu} & \frac{\mathbf{A}_{23}^T}{\mu} & \frac{\mathbf{A}_{33}^T}{\mu} \end{bmatrix} \begin{bmatrix} \mathbf{q}_1 \\ \mathbf{q}_2 \\ \mathbf{q}_3 \end{bmatrix} + \begin{bmatrix} \mathbf{C}_1^T \\ \frac{\mathbf{C}_2^T}{\varepsilon} \\ \frac{\mathbf{C}_3^T}{\mu} \end{bmatrix} \nu. \tag{7.40}$$

Later, the system (7.40) is structured to a lower triangular form

$$\begin{bmatrix} \dot{\mathbf{q}}_s \\ \dot{\mathbf{q}}_{fs} \\ \dot{\mathbf{q}}_3 \end{bmatrix} = \begin{bmatrix} \mathbf{A}_s^T & \mathbf{0} & \mathbf{0} \\ \frac{\bar{\mathbf{A}}_{12}^T}{\varepsilon} & \frac{\mathbf{A}_{f1}^T}{\varepsilon} & \mathbf{0} \\ \frac{\mathbf{A}_{13}^T}{\mu} & \frac{\bar{\mathbf{A}}_{23}^T}{\mu} & \frac{\mathbf{A}_{f2}^T}{\mu} \end{bmatrix} \begin{bmatrix} \mathbf{q}_s \\ \mathbf{q}_{fs} \\ \mathbf{q}_3 \end{bmatrix} + \begin{bmatrix} \mathbf{C}_s^T \\ \frac{\bar{\mathbf{C}}_{f1}^T}{\varepsilon} \\ \frac{\mathbf{C}_3^T}{\mu} \end{bmatrix} \nu, \tag{7.41}$$

by applying state transformation

$$\begin{bmatrix} \mathbf{q}_s \\ \mathbf{q}_{fs} \\ \mathbf{q}_3 \end{bmatrix} = \begin{bmatrix} \mathbf{I}_{n1} & \varepsilon\mathbf{P}_{21}^T & \mu\mathbf{P}_{31}^T \\ \mathbf{0} & \mathbf{I}_{n2} & \mu\mathbf{P}_{32}^T \\ \mathbf{0} & \mathbf{0} & \mathbf{I}_{n3} \end{bmatrix} \begin{bmatrix} \mathbf{q}_1 \\ \mathbf{q}_2 \\ \mathbf{q}_3 \end{bmatrix} = \mathbf{T}_2^T \begin{bmatrix} \mathbf{q}_1 \\ \mathbf{q}_2 \\ \mathbf{q}_3 \end{bmatrix}, \tag{7.42}$$

to (7.40). Herein, $\mathbf{P}_{21}^T$ and $\mathbf{P}_{31}^T$ are reckoned by solving

$$\left.\begin{aligned} -\varepsilon\mathbf{A}_s^T\mathbf{P}_{21}^T + \mathbf{A}_{21}^T + \mathbf{P}_{21}^T\mathbf{A}_{22}^T + \mathbf{P}_{31}^T\mathbf{A}_{23}^T = 0, \\ -\mu\mathbf{A}_s^T\mathbf{P}_{31}^T + \mathbf{A}_{31}^T + \mathbf{P}_{21}^T\mathbf{A}_{32}^T + \mathbf{P}_{31}^T\mathbf{A}_{33}^T = 0, \end{aligned}\right\} \tag{7.43}$$

iteratively, in the same way as that of (7.9). After that, one can gain $\mathbf{A}_s^T = \mathbf{A}_{11}^T + \mathbf{P}_{21}^T\mathbf{A}_{12}^T + \mathbf{P}_{31}^T\mathbf{A}_{13}^T$, $\mathbf{A}_{f22}^T = -\varepsilon\mathbf{A}_{12}^T\mathbf{P}_{21}^T + \mathbf{A}_{22}^T$, $\mathbf{A}_{f32}^T = -\mu\mathbf{A}_{12}^T\mathbf{P}_{31}^T + \mathbf{A}_{32}^T$, $\bar{\mathbf{A}}_{23}^T = -\varepsilon\mathbf{A}_{13}^T\mathbf{P}_{21}^T + \mathbf{A}_{23}^T$, $\mathbf{A}_{f33}^T = -\mu\mathbf{A}_{13}^T\mathbf{P}_{31}^T + \mathbf{A}_{33}^T$ and $\mathbf{C}_s^T = \mathbf{C}_1^T + \mathbf{P}_{21}^T\mathbf{C}_2^T + \mathbf{P}_{31}^T\mathbf{C}_3^T$. Again, $\mathbf{P}_{32}^T$ is determined using

$$-\mu(\mathbf{A}_{f22}^T + \varepsilon\mathbf{P}_{32}^T\bar{\mathbf{A}}_{23}^T)\mathbf{P}_{32}^T + (\mathbf{A}_{f32}^T + \mathbf{P}_{32}^T\mathbf{A}_{f33}^T) = 0, \tag{7.44}$$

with Assumption 7.2 for a small value of $\mu$. Subsequently, one can have $\bar{\mathbf{A}}_{12}^T = \mathbf{A}_{12}^T + \varepsilon\mathbf{P}_{32}^T\mathbf{A}_{13}^T$, $\mathbf{A}_{f1}^T = \mathbf{A}_{f22}^T + \varepsilon\mathbf{P}_{32}^T\bar{\mathbf{A}}_{23}^T$, $\bar{\mathbf{A}}_{f2}^T = -\mu\bar{\mathbf{A}}_{23}^T\mathbf{P}_{32}^T + \mathbf{A}_{f33}^T$ and $\bar{\mathbf{C}}_{f1}^T = \varepsilon\mathbf{P}_{32}^T\mathbf{C}_3^T + \mathbf{C}_2^T$. Since, $\psi(\mathbf{A}) = \psi(\mathbf{A}^T)$, the system (7.41) also conforms to the eigenvalue grouping property of the system (7.7). For the three-stage observer design, in the first stage, applying

$$\nu = -\mathbf{L}_s^T\mathbf{q}_s + \nu_{f1}, \tag{7.45}$$

to the system (7.41), leads to

$$\begin{bmatrix} \dot{\mathbf{q}}_s \\ \dot{\mathbf{q}}_{fs} \\ \dot{\mathbf{q}}_3 \end{bmatrix} = \begin{bmatrix} \mathbf{A}_s^T - \mathbf{C}_s^T\mathbf{L}_s^T & \mathbf{0} & \mathbf{0} \\ \frac{\bar{\mathbf{A}}_{12}^T - \bar{\mathbf{C}}_{f1}^T\mathbf{L}_s^T}{\varepsilon} & \frac{\mathbf{A}_{f1}^T}{\varepsilon} & \mathbf{0} \\ \frac{\mathbf{A}_{13}^T - \mathbf{C}_3^T\mathbf{L}_s^T}{\mu} & \frac{\bar{\mathbf{A}}_{23}^T}{\mu} & \frac{\mathbf{A}_{f2}^T}{\mu} \end{bmatrix} \begin{bmatrix} \mathbf{q}_s \\ \mathbf{q}_{fs} \\ \mathbf{q}_3 \end{bmatrix} + \begin{bmatrix} \mathbf{C}_s^T \\ \frac{\bar{\mathbf{C}}_{f1}^T}{\varepsilon} \\ \frac{\mathbf{C}_3^T}{\mu} \end{bmatrix} \nu_{f1}, \tag{7.46}$$

so that $\psi(\mathbf{A}_s^T - \mathbf{C}_s^T\mathbf{L}_s^T) = \psi_s^{desired}$. As the transformations (7.39) and (7.42) transmutes the original system from (7.38) to (7.41) without the loss of controllability (or observability), the pair $(\mathbf{A}_s^T, \mathbf{C}_s^T)$ is controllable (or the pair $(\mathbf{A}_s, \mathbf{C}_s)$ is observable), due to which an arbitrary pole assignment is possible under Assumption 7.1. Next, the system (7.46) is restructured as

$$\begin{bmatrix} \dot{\mathbf{q}}_s \\ \dot{\mathbf{q}}_{f1} \\ \dot{\mathbf{q}}_3 \end{bmatrix} = \begin{bmatrix} \mathbf{A}_s^T - \mathbf{C}_s^T\mathbf{L}_s^T & \mathbf{0} & \mathbf{0} \\ \mathbf{0} & \frac{\mathbf{A}_{f1}^T}{\varepsilon} & \mathbf{0} \\ \frac{\bar{\mathbf{A}}_{13}^T}{\mu} & \frac{\bar{\mathbf{A}}_{23}^T}{\mu} & \frac{\mathbf{A}_{f2}^T}{\mu} \end{bmatrix} \begin{bmatrix} \mathbf{q}_s \\ \mathbf{q}_{f1} \\ \mathbf{q}_3 \end{bmatrix} + \begin{bmatrix} \mathbf{C}_s^T \\ \frac{\mathbf{C}_{f1}^T}{\varepsilon} \\ \frac{\mathbf{C}_3^T}{\mu} \end{bmatrix} \nu_{f1}, \tag{7.47}$$

by applying state variable transformation

$$\begin{bmatrix} \mathbf{q}_s \\ \mathbf{q}_{f1} \\ \mathbf{q}_3 \end{bmatrix} = \begin{bmatrix} \mathbf{I}_{n1} & \mathbf{0} & \mathbf{0} \\ \mathbf{P}_{12}^T & \mathbf{I}_{n2} & \mathbf{0} \\ \mathbf{0} & \mathbf{0} & \mathbf{I}_{n3} \end{bmatrix} \begin{bmatrix} \mathbf{q}_s \\ \mathbf{q}_{fs} \\ \mathbf{q}_3 \end{bmatrix} = \mathbf{T}_3^T \begin{bmatrix} \mathbf{q}_s \\ \mathbf{q}_{fs} \\ \mathbf{q}_3 \end{bmatrix}, \tag{7.48}$$

to the system (7.46), in which $\bar{\mathbf{A}}_{13}^T = \mathbf{A}_{13}^T - \mathbf{C}_3^T\mathbf{L}_s^T - \bar{\mathbf{A}}_{23}^T\mathbf{P}_{12}^T$, $\mathbf{C}_{f1}^T = \bar{\mathbf{C}}_{f1}^T + \varepsilon\mathbf{P}_{12}^T\bar{\mathbf{C}}_s^T$ and

$$\varepsilon\mathbf{P}_{12}^T(\mathbf{A}_s^T - \mathbf{C}_s^T\mathbf{L}_s^T) + \bar{\mathbf{A}}_{12}^T - \bar{\mathbf{C}}_{f1}^T\mathbf{L}_s^T - \mathbf{A}_{f1}^T\mathbf{P}_{12}^T = 0. \tag{7.49}$$

Just like (7.18), (7.49) can be solved to get $\mathbf{P}_{12}^T$ with:

**Assumption 7.8** $\psi(\mathbf{A}_s^T - \mathbf{C}_s^T\mathbf{L}_s^T) \neq \psi\left(\frac{\mathbf{A}_{f1}^T}{\varepsilon}\right)$.

Thereupon, in the second stage, passing input

$$\nu_{f1} = -\mathbf{L}_{f1}^T\mathbf{q}_{f1} + \nu_{f2}, \tag{7.50}$$

to the system (7.47) provides

$$\begin{bmatrix} \dot{\mathbf{q}}_s \\ \dot{\mathbf{q}}_{f1} \\ \dot{\mathbf{q}}_3 \end{bmatrix} = \begin{bmatrix} \mathbf{A}_s^T - \mathbf{C}_s^T\mathbf{L}_s^T & -\mathbf{C}_s^T\mathbf{L}_{f1}^T & \mathbf{0} \\ \mathbf{0} & \frac{\mathbf{A}_{f1}^T - \mathbf{C}_{f1}^T\mathbf{L}_{f1}^T}{\varepsilon} & \mathbf{0} \\ \frac{\bar{\mathbf{A}}_{13}^T}{\mu} & \frac{\bar{\mathbf{A}}_{23}^T - \mathbf{C}_3^T\mathbf{L}_{f1}^T}{\mu} & \frac{\mathbf{A}_{f2}^T}{\mu} \end{bmatrix} \begin{bmatrix} \mathbf{q}_s \\ \mathbf{q}_{f1} \\ \mathbf{q}_3 \end{bmatrix} + \begin{bmatrix} \mathbf{C}_s^T \\ \frac{\mathbf{C}_{f1}^T}{\varepsilon} \\ \frac{\mathbf{C}_3^T}{\mu} \end{bmatrix} \nu_{f2}, \tag{7.51}$$

so as to place $\psi\left(\frac{\mathbf{A}_{f1}^T - \mathbf{C}_{f1}^T\mathbf{L}_{f1}^T}{\varepsilon}\right) = \psi_{f1}^{desired}$. Controllability of $\left(\frac{\mathbf{A}_{f1}^T}{\varepsilon}, \frac{\mathbf{C}_{f1}^T}{\varepsilon}\right)$ is preserved due to Assumption 7.1. Further, very fast subsystem is uncoupled from slow and fast subsystems

$$\begin{bmatrix} \dot{\mathbf{q}}_s \\ \dot{\mathbf{q}}_{f1} \\ \dot{\mathbf{q}}_{f2} \end{bmatrix} = \begin{bmatrix} \mathbf{A}_s^T - \mathbf{C}_s^T \mathbf{L}_s^T & -\mathbf{C}_s^T \mathbf{L}_{f1}^T & \mathbf{0} \\ \mathbf{0} & \frac{\mathbf{A}_{f1}^T - \mathbf{C}_{f1}^T \mathbf{L}_{f1}^T}{\varepsilon} & \mathbf{0} \\ \mathbf{0} & \mathbf{0} & \frac{\mathbf{A}_{f2}^T}{\mu} \end{bmatrix} \begin{bmatrix} \mathbf{q}_s \\ \mathbf{q}_{f1} \\ \mathbf{q}_{f2} \end{bmatrix} + \begin{bmatrix} \mathbf{C}_s^T \\ \frac{\mathbf{C}_{f1}^T}{\varepsilon} \\ \frac{\mathbf{C}_{f2}^T}{\mu} \end{bmatrix} \nu_{f2}, \tag{7.52}$$

by applying linear state transformation

$$\begin{bmatrix} \mathbf{q}_s \\ \mathbf{q}_{f1} \\ \mathbf{q}_{f2} \end{bmatrix} = \begin{bmatrix} \mathbf{I}_{n1} & \mathbf{0} & \mathbf{0} \\ \mathbf{0} & \mathbf{I}_{n2} & \mathbf{0} \\ \mathbf{P}_{13}^T & \mathbf{P}_{23}^T & \mathbf{I}_{n3} \end{bmatrix} \begin{bmatrix} \mathbf{q}_s \\ \mathbf{q}_{f1} \\ \mathbf{q}_3 \end{bmatrix} = \mathbf{T}_4^T \begin{bmatrix} \mathbf{q}_s \\ \mathbf{q}_{f1} \\ \mathbf{q}_3 \end{bmatrix}, \tag{7.53}$$

to (7.51), where $\mathbf{C}_{f2}^T = \mu \mathbf{P}_{13}^T \mathbf{C}_s^T + \frac{\mu}{\varepsilon} \mathbf{P}_{23}^T \mathbf{C}_{f1}^T + \mathbf{C}_3^T$ and $\mathbf{P}_{13}^T$ and $\mathbf{P}_{23}^T$ are estimated by computing

$$\mu \mathbf{P}_{13}^T (\mathbf{A}_s^T - \mathbf{C}_s^T \mathbf{L}_s^T) + \bar{\mathbf{A}}_{13}^T - \mathbf{A}_{f2}^T \mathbf{P}_{13}^T = 0, \tag{7.54}$$

$$-\mu \mathbf{P}_{13}^T \mathbf{C}_s^T \mathbf{L}_{f1}^T + \frac{\mu}{\varepsilon} \mathbf{P}_{23}^T (\mathbf{A}_{f1}^T - \mathbf{C}_{f1}^T \mathbf{L}_{f1}^T) + \bar{\mathbf{A}}_{23} - \mathbf{C}_3^T \mathbf{L}_{f1}^T - \bar{\mathbf{A}}_{f2}^T \mathbf{P}_{23}^T = 0, \tag{7.55}$$

using the 'lyap' command with the following assumptions:

**Assumption 7.9** $\psi(\mathbf{A}_s^T - \mathbf{C}_s^T \mathbf{L}_s^T) \neq \psi\left(\frac{\mathbf{A}_{f2}^T}{\mu}\right)$ for (7.54).

**Assumption 7.10** $\psi\left(\frac{\mathbf{A}_{f1}^T - \mathbf{C}_{f1}^T \mathbf{L}_{f1}^T}{\varepsilon}\right) \neq \psi\left(\frac{\mathbf{A}_{f2}^T}{\mu}\right)$ for (7.55).

Systems $(\mathbf{A}_s^T - \mathbf{C}_s^T \mathbf{L}_s^T)$ and $\left(\frac{\mathbf{A}_{f1}^T - \mathbf{C}_{f1}^T \mathbf{L}_{f1}^T}{\varepsilon}\right)$ are user designed with stable eigenvalues laying in slow and fast regions respectively and $\frac{\mathbf{A}_{f2}^T}{\mu}$ has very fast eigenvalues of original system. That being so, Assumptions 7.9 and 7.10 are complied. Finally,

$$\nu_{f2} = -\mathbf{L}_{f2}^T \mathbf{q}_{f2}, \tag{7.56}$$

is applied to the system (7.52), in the third stage, as

$$\begin{bmatrix} \dot{\mathbf{q}}_s \\ \dot{\mathbf{q}}_{f1} \\ \dot{\mathbf{q}}_{f2} \end{bmatrix} = \begin{bmatrix} \mathbf{A}_s^T - \mathbf{C}_s^T \mathbf{L}_s^T & -\mathbf{C}_s^T \mathbf{L}_{f1}^T & -\mathbf{C}_s^T \mathbf{L}_{f2}^T \\ \mathbf{0} & \frac{\mathbf{A}_{f1}^T - \mathbf{C}_{f1}^T \mathbf{L}_{f1}^T}{\varepsilon} & -\mathbf{C}_{f1}^T \mathbf{L}_{f2}^T \\ \mathbf{0} & \mathbf{0} & \frac{\mathbf{A}_{f2}^T - \mathbf{C}_{f2}^T \mathbf{L}_{f2}^T}{\mu} \end{bmatrix} \begin{bmatrix} \mathbf{q}_s \\ \mathbf{q}_{f1} \\ \mathbf{q}_{f2} \end{bmatrix}, \tag{7.57}$$

with the aim to assign $\psi\left(\frac{\mathbf{A}_{f2}^T - \mathbf{C}_{f2}^T \mathbf{L}_{f2}^T}{\mu}\right) = \psi_{f2}^{desired}$, assuming the controllability of $\left(\frac{\mathbf{A}_{f2}^T}{\mu}, \frac{\mathbf{C}_{f2}^T}{\mu}\right)$ is satisfied by Assumption 7.1. The overall observer gain is given by

$$\nu = \mathbf{L}^T \mathbf{x} = \left[\mathbf{L}_1^T \ \frac{1}{\varepsilon} \mathbf{L}_2^T \ \frac{1}{\mu} \mathbf{L}_3^T\right] \left[\mathbf{x}_1^T \ \mathbf{x}_2^T \ \mathbf{x}_3^T\right]^T, \tag{7.58}$$

where $\mathbf{L}_1^T = \mathbf{L}_s^T + \mathbf{L}_f^T\mathbf{P}_{13}^T + (\mathbf{L}_{f2}^T\mathbf{P}_{23}^T + \mathbf{L}_{f1}^T)\mathbf{P}_{12}^T$, $\mathbf{L}_2^T = \varepsilon\mathbf{L}_1^T\mathbf{P}_{21}^T + \mathbf{L}_{f2}^T\mathbf{P}_{23}^T + \mathbf{L}_{f1}^T$, and $\mathbf{L}_3^T = \mu\mathbf{L}_1^T\mathbf{P}_{31}^T + \mu(\mathbf{L}_{f2}^T\mathbf{P}_{23}^T + \mathbf{L}_{f1}^T)\mathbf{P}_{32}^T + \mathbf{L}_{f2}^T$. Once again, applying the change of states as

$$\left[\mathbf{x}_s^T \ \mathbf{x}_{f1}^T \ \mathbf{x}_{f2}^T\right]^T = \mathbf{T}_1^{-T}\left[\mathbf{q}_s^T \ \mathbf{q}_{f1}^T \ \mathbf{q}_{f2}^T\right]^T, \tag{7.59}$$

to the system (7.57) results

$$\begin{bmatrix} \dot{\mathbf{x}}_s \\ \dot{\mathbf{x}}_{f1} \\ \dot{\mathbf{x}}_{f2} \end{bmatrix} = \begin{bmatrix} (\mathbf{A}_s - \mathbf{L}_s\mathbf{C}_s)^T & \frac{-(\mathbf{L}_{f1}\mathbf{C}_s)^T}{\varepsilon} & \frac{-(\mathbf{L}_{f2}\mathbf{C}_s)^T}{\mu} \\ \mathbf{0} & \frac{(\mathbf{A}_{f1}-\mathbf{L}_{f1}\mathbf{C}_{f1})^T}{\varepsilon} & \frac{-(\mathbf{L}_{f2}\mathbf{C}_{f1})^T}{\mu} \\ \mathbf{0} & \mathbf{0} & \frac{(\mathbf{A}_{f2}-\mathbf{L}_{f2}\mathbf{C}_{f2})^T}{\mu} \end{bmatrix} \begin{bmatrix} \mathbf{x}_s \\ \mathbf{x}_{f1} \\ \mathbf{x}_{f2} \end{bmatrix}. \tag{7.60}$$

The hypothetical system (7.38) and system (7.60) are related as

$$\left.\begin{aligned} \left[\mathbf{x}_s^T \ \mathbf{x}_{f1}^T \ \mathbf{x}_{f2}^T\right]^T &= \mathbf{T}^T\left[\mathbf{x}_1^T \ \mathbf{x}_2^T \ \mathbf{x}_3^T\right]^T, \\ \left[\mathbf{x}_1^T \ \mathbf{x}_2^T \ \mathbf{x}_3^T\right]^T &= \mathbf{T}^{-T}\left[\mathbf{x}_s^T \ \mathbf{x}_{f1}^T \ \mathbf{x}_{f2}^T\right]^T, \end{aligned}\right\} \tag{7.61}$$

where $\mathbf{T}^T = \mathbf{T}_1^{-T}\mathbf{T}_4^T\mathbf{T}_3^T\mathbf{T}_2^T\mathbf{T}_1^T$. Using (7.60) and the duality principle (Chen 1999; Yoo and Gajic 2018), observer can be configured as

$$\begin{bmatrix} \dot{\hat{\mathbf{z}}}_s \\ \dot{\hat{\mathbf{z}}}_{f1} \\ \dot{\hat{\mathbf{z}}}_{f2} \end{bmatrix} = \begin{bmatrix} (\mathbf{A}_s - \mathbf{L}_s\mathbf{C}_s) & \mathbf{0} & \mathbf{0} \\ \frac{-(\mathbf{L}_{f1}\mathbf{C}_s)}{\varepsilon} & \frac{(\mathbf{A}_{f1}-\mathbf{L}_{f1}\mathbf{C}_{f1})}{\varepsilon} & \mathbf{0} \\ \frac{-(\mathbf{L}_{f2}\mathbf{C}_s)}{\mu} & \frac{-(\mathbf{L}_{f2}\mathbf{C}_{f1})}{\mu} & \frac{(\mathbf{A}_{f2}-\mathbf{L}_{f2}\mathbf{C}_{f2})}{\mu} \end{bmatrix} \begin{bmatrix} \hat{\mathbf{z}}_s \\ \hat{\mathbf{z}}_{f1} \\ \hat{\mathbf{z}}_{f2} \end{bmatrix} + \begin{bmatrix} \mathbf{B}_s \\ \frac{\mathbf{B}_{f1}}{\varepsilon} \\ \frac{\mathbf{B}_{f2}}{\mu} \end{bmatrix}\mathbf{u} + \begin{bmatrix} \mathbf{L}_s \\ \frac{\mathbf{L}_{f1}}{\varepsilon} \\ \frac{\mathbf{L}_{f2}}{\mu} \end{bmatrix}\mathbf{y}. \tag{7.62}$$

Employing (7.61) for three-time-scale system (7.36)–(7.37) gives

$$\left[\hat{\mathbf{z}}_1^T \ \hat{\mathbf{z}}_2^T \ \hat{\mathbf{z}}_3^T\right]^T = \mathbf{T}\left[\hat{\mathbf{z}}_s^T \ \hat{\mathbf{z}}_{f1}^T \ \hat{\mathbf{z}}_{f2}^T\right]^T, \tag{7.63}$$

resulting into

$$\begin{bmatrix} \dot{\hat{\mathbf{z}}}_s \\ \dot{\hat{\mathbf{z}}}_{f1} \\ \dot{\hat{\mathbf{z}}}_{f2} \end{bmatrix} = \mathbf{T}^{-1}(\mathbf{A} - \mathbf{L}\mathbf{C})\mathbf{T}\begin{bmatrix} \hat{\mathbf{z}}_s \\ \hat{\mathbf{z}}_{f1} \\ \hat{\mathbf{z}}_{f2} \end{bmatrix} + \mathbf{T}^{-1}\mathbf{B}\mathbf{u} + \mathbf{T}^{-1}\mathbf{L}\mathbf{y}. \tag{7.64}$$

Systems (7.62) and (7.64) are equivalent, in which case

$$\mathbf{T}^{-1}\mathbf{L} = \left[\mathbf{L}_s^T \ \frac{\mathbf{L}_{f1}^T}{\varepsilon} \ \frac{\mathbf{L}_{f2}^T}{\mu}\right]^T \Longrightarrow \mathbf{L} = \mathbf{T}\left[\mathbf{L}_s^T \ \frac{\mathbf{L}_{f1}^T}{\varepsilon} \ \frac{\mathbf{L}_{f2}^T}{\mu}\right]^T,$$

can be verified with $\mathbf{L}$ in (7.58) and $\left[\mathbf{B}_s^T \ \frac{\mathbf{B}_{f1}^T}{\varepsilon} \ \frac{\mathbf{B}_{f2}^T}{\mu}\right]^T$ can be calculated using $\mathbf{T}^{-1}\mathbf{B}$.

**Remark 7.1** As the designs of MRSF control and sequential observer are done by using suggested block-triangular forms of the system, instead of block-diagonal forms (Shimjith et al. 2013), less number of similarity transformations are needed and less number of algebraic equations are required to be solved. This reduced both offline and online computational efforts.

## 7.3 Numerical Example

Consider a ninth-order system (Jamshidi 1983), having 3 slow, 2 fast, and 4 very fast eigenvalues as

$$\mathbf{A} = \begin{bmatrix} -0.311 & 0 & 1.028 & 0 & 0 & 2 & 0.10 & -6.30 & -0.40 \\ 1.01 & -0.65 & -0.084 & 0 & 0.83 & 1.70 & 3 & 2.40 & 0.20 \\ -0.11 & 0 & -1.26 & 0 & -6.30 & -1.04 & 0 & 4.10 & 0.01 \\ 0 & 0.10 & 0 & -13.00 & 0 & 0.48 & 0 & 0 & 0 \\ 0 & -0.05 & 0.25 & 0 & -17.60 & -0.30 & 0 & 0 & 0 \\ 0 & 4.50 & -3.80 & -640 & 375 & -97.50 & 1.12 & 0 & -0.80 \\ 0 & -2.90 & 2.75 & 11.10 & 0 & 1.01 & -143 & 0.50 & 0 \\ 2.50 & -5.70 & -4.00 & 0 & 0 & -0.91 & 0 & -184 & -4.20 \\ 1.80 & -3.10 & 3.18 & 0 & 1.76 & -0.70 & 1.53 & 3.20 & -207 \end{bmatrix},$$

$$\mathbf{B} = \begin{bmatrix} 0.8812 \; 1.1366 \; -0.6790 \; -0.1311 \; -0.2270 \; 0.9823 \; -1.0174 \; -0.8227 \; 0.7900 \end{bmatrix}^T,$$
$$\mathbf{C} = \begin{bmatrix} 1 \; 1 \; 1 \; 1 \; 1 \; 1 \; 1 \; 1 \; 1 \end{bmatrix}.$$

Eigenvalues are located at (−206.40, −184.35, −142.96, −92.26, −20.38, −15.31, −1.56, −0.80, −0.30) with

$$\varepsilon = \frac{max|\psi(\mathbf{A}_s)|}{min\left|\psi\left(\frac{\mathbf{A}_{f1}}{\varepsilon}\right)\right|} = 0.1021, \mu = \frac{max|\psi(\mathbf{A}_s)|}{min\left|\psi\left(\frac{\mathbf{A}_{f2}}{\mu}\right)\right|} = 0.0169.$$

Thus, $N_1$ and $N_2$ are taken as 4 and 5 respectively. If $\tau = 2$ s, then $\Delta = 0.5$ s and $\delta = 0.1$ s. Suppose the desired eigenvalues of the continuous-time system are (−180, −150, −130, −110, −18, −16, −2.6, −1.8, −0.8). Consequently, the desired eigenvalues in $s$-plane for the slow, fast and very fast subsystems for $\tau$, $\Delta$ and $\delta$ sampling intervals respectively can be obtained by $e^{\psi_i T}$, where $\psi_i$ is the $i$th eigenvalue of the continuous-time system and $T$ is the sampling period. Now, solutions of (7.9) and (7.10) are found by fixed point iteration method (Naidu 1988) and an upper triangular form (7.7) is reached. This system is then discretized with a sampling period of 0.1 s. Later, to place $\psi_{\delta f2}^{desired}$, the feedback gain is determined as

$$\mathbf{F}_{f2} = \begin{bmatrix} 0.0079 \; 0.0002 \; -0.0000 \; -0.0001 \end{bmatrix}.$$

Next, $\mathbf{M}_{23}$ is estimated to get (7.16) and then represented as (7.19) with $\Delta$ sampling interval using (7.20). The fast subsystem eigenvalues are placed at $\psi_{\Delta f1}^{desired}$ by computing

$$\mathbf{F}_{f1} = \begin{bmatrix} -0.0118 & 0.0049 \end{bmatrix}.$$

Finally, $\mathbf{M}_{12}$ and $\mathbf{M}_{13}$ are determined to get (7.23) and then represented into (7.27) with $\tau$ sampling interval using (7.28). The feedback gain

$$\mathbf{F}_s = \begin{bmatrix} 0.9350 & 0.1143 & 0.9972 \end{bmatrix},$$

is computed to place placed $\psi_{\tau s}^{desired}$. Utilizing obtained feedback gains, control input (7.31) is formulated. System initial conditions are taken as (0.1, 0, 0, −0.2, 0, 0.05, 0, 0, 0). States $\hat{\mathbf{z}}_s^T$, $\hat{\mathbf{z}}_{f1}^T$ and $\hat{\mathbf{z}}_{f2}^T$ are estimated using the sequential observer and are sampled at $\tau$, $\Delta$ and $\delta$ sampling intervals respectively. Desired observer eigenvalues are taken as (−540, −450, −390, −310, −54, −48, −7.8, −5.4, −2.7). For this, solutions of (7.43) and (7.44) are determined and system (7.41) is obtained. After this, to position $\psi_s^{desired}$, the observer gain is determined as

$$\mathbf{L}_s^T = \begin{bmatrix} 27.0435 & -58.0586 & 37.5108 \end{bmatrix}.$$

Further, $\mathbf{P}_{12}^T$ is found and system (7.47) is formed. The fast observer eigenvalues are placed at $\psi_{f1}^{desired}$ by computing

$$\mathbf{L}_{f1}^T = \begin{bmatrix} -3.1335 & -3.2063 \end{bmatrix}.$$

At the end, $\mathbf{P}_{13}^T$ and $\mathbf{P}_{23}^T$ are calculated to gain (7.52). The very fast subsystem eigenvalues are placed at $\psi_{f2}^{desired}$ by determining

$$\mathbf{L}_{f2}^T = \begin{bmatrix} 96.74 & -357.93 & 279.31 & -71.05 \end{bmatrix}.$$

Using all these feedback and observer gains, the system is simulated. The simulation results are compared with the single-rate sampling viz. $\delta$ and $\Delta$ seconds. Figure 7.1 shows the evolution of original states, and Fig. 7.2 depicts the estimation of decoupled states obtained by sampling them with respective sampling periods. Initial conditions for the estimated states are obtained from (7.61) (Yoo and Gajic 2018). From Figs. 7.1 and 7.2 improvement in the overall system performance can be observed with MRSF control.

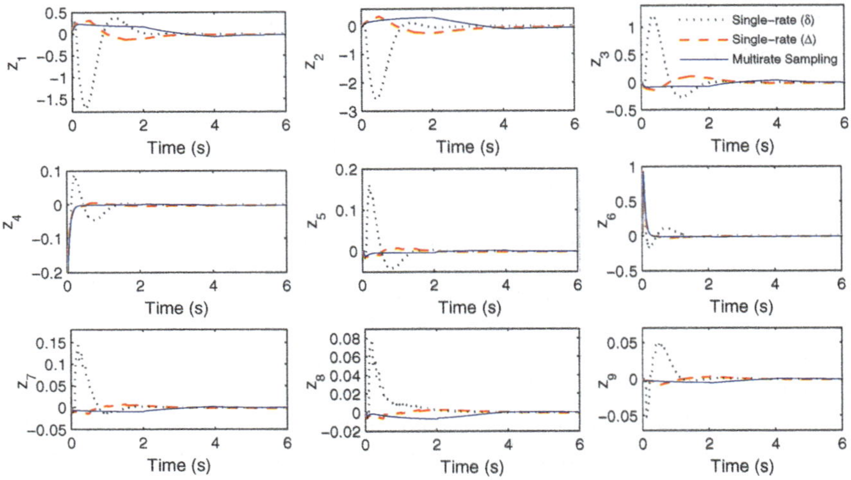


**Fig. 7.1** Evolution of states of the original three-time-scale system

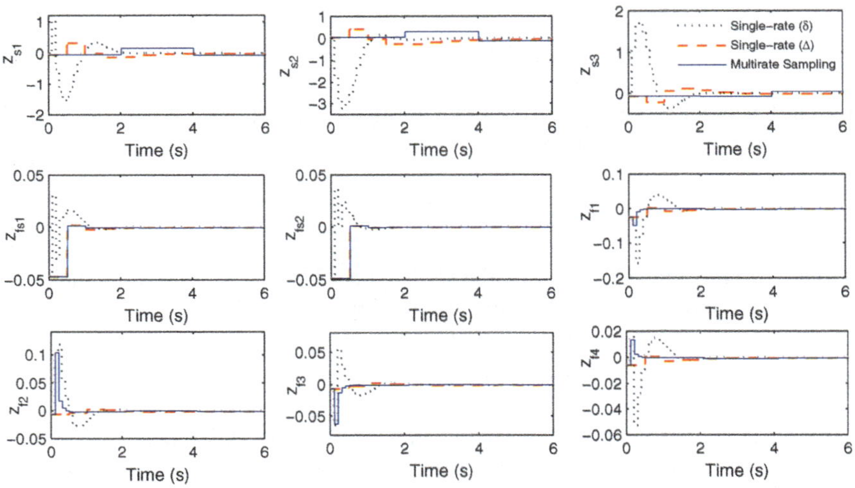


**Fig. 7.2** Evolution of the estimated slow ($\hat{\mathbf{z}}_s$), fast ($\hat{\mathbf{z}}_{fs}$) and very fast ($\hat{\mathbf{z}}_f$) states after sampling

## 7.4 Conclusion

In this chapter, multirate state feedback control is presented for a three-time-scale system using the exact separation of the very fast, fast, and slow subsystems. The very fast subsystem states are sampled at a higher rate than the fast subsystem states and fast subsystem states are sampled at a higher rate than slow subsystem states. The relation between the sampling periods of very fast, fast, and slow subsystems is obtained. Three steps of subsystem separation and three stages of MRSF control are merged together. For control implementation, the decoupled states are computed by a sequential three-stage observer, designed for a continuous three-time-scale system. The proposed control is then applied to an illustrative example and simulations are presented. Simulations are compared with the single-rate sampling and it is observed that the multirate sampling actually improves system performance.

## References

Chen CT (1999) Linear system theory and design. Oxford University Press, New York
Jamshidi M (1983) Large-scale systems: modeling and control. Elsevier, New York
Kokotovic PV, Khalil HK, O'Reilly J (1999) Singular perturbation methods in control: analysis and design. SIAM Publishers, Philadelphia
MATLAB/Simulink Control Design Toolbox User Manual. MathWorks, Version 4.5 (2017)
Munje R, Gu D, Desai R, Patre B, Zhang W (2018) Observer-based spatial control of advanced heavy water reactor using time-scale decoupling. IEEE Trans Nucl Sci 65:2756–2766
Naidu DS (1988) Singular perturbation methodology in control systems. Peter Peregrinus Ltd, London U.K
Radisavljevic-Gajic V, Milanovic M, Clayton G (2017) Three-stage feedback controller design with application to three-time-scale linear control system. ASME J Dyn Syst, Meas, Control 139:1–10
Shimjith SR, Tiwari AP, Bandyopadhyay B (2013) Direct block diagonalization and composite control of three-time-scale systems. In: Modeling and control of a large nuclear reactor, pp 101–114. Springer Berlin Heidelberg
Yoo H, Gajic Z (2018) New designs of linear observers and observer-based controllers for singularly perturbed linear systems. IEEE Trans Auto Control 63:3904–3911

# Chapter 8
# Concluding Remarks and Future Directions

**Abstract** In this chapter, all the multi-stage feedback control schemes suggested in this monograph are summarized and their benefits are listed for quick grasp. Also, the chapter highlights the future areas in the multi-stage feedback control designs both in continuous and discrete-time domains.

**Keywords** Continuous-time · Discrete-time · Feedback control · Multirate sampling · Sequential observer · Three-stage design · Three-time-scale system · Singular perturbation

## 8.1 Concluding Remarks

Generally, the physical systems having combinations of components from electrical, mechanical, chemical, thermodynamic, electrochemical, etc. systems, represent the multi-time-scale property (Kokotovic 1976; Syrcos and Sannuti 1983; Saksena et al. 1984; Naidu et al. 1987; Naidu 2002; Zhang et al. 2014). Some of the important examples discussed in the literature are nuclear reactor (Shimjith et al. 2013; Munje et al. 2018), proton-exchange membrane fuel cell (Radisavljević-Gajić et al. 2019), thermal power plants (Naidu and Rao 1985; Gajic and Lim 2001; Radisavljevic-Gajic 2015), power system (Jamshidi 1983), wind energy (Nguyen and Naidu 2012), etc. Multi-stage designs of feedback controller and full- or reduced-order observer for such a system help to overcome ill-conditioning of system matrices arising due to multi-time scale. In this monograph, the three-stage feedback controller design, two-stage (reduced-order), and three-stage (full-order) observer designs for the continuous three-time-scale system are suggested. This monograph also presents, two-stage and three-stage multirate state feedback control designs for the discrete two-time-scale and three-time-scale systems, respectively, in which the decoupled states are estimated by formulating two-stage and three-stage sequential (full-order) observers. The applicability of these methods to physical systems and numerical examples is established. A comprehensive chapter-wise summary of all the designed methods is given in Table 8.1. The overall benefits of the designed controllers and observers are as follows.



R. Munje et al., *Multi-stage Feedback Control Design for Multi-time-scale System*, Springer Tracts in Electrical and Electronics Engineering,
https://doi.org/10.1007/978-981-97-4592-0_8

**Table 8.1** Design summary of the control schemes

| Chapter number | Type of design[a] | Feedback control | Observer design[b] | Considered system | Remark |
|---|---|---|---|---|---|
| 2 | Continuous-time | Three-stage | ND[c] | Linear system with three groups of eigenvalues | No consideration of singular perturbation parameters |
| 3 | Continuous-time | Three-stage | Three-stage | Three-time-scale system | Feedback control for the system with integral action and full-order observer |
| 4 | Continuous-time | ND | Two-stage, Three-stage | Three-time-scale system | Full- and reduced-order observer designs |
| 5 | Discrete-time | Two-stage | ND | Two-time-scale system | Two schemes of multirate feedback control depending on sampling sequence |
| 6 | Discrete-time | Two-stage | Two-stage | Two-time-scale system | Multirate state feedback control with state estimation by sequential observer |
| 7 | Discrete-time | Three-stage | Three-stage | Three-time-scale system | Multirate state feedback control with state estimation by sequential observer |

[a]It is only applicable for the feedback control design
[b]All the observers are designed in a continuous-time domain
[c]ND—Not designed

1. Simplicity in the manual three-stage designs of feedback controller and full-order observer due to (i) the use of the two-time-scale ($2 \times 2$) formulation instead of three-time-scale ($3 \times 3$) formulation and (ii) less number of state variable transformation matrices for a three-time-scale system.
2. Reduction in offline computation as less number of algebraic equations are required to be solved.
3. Accommodation of multirate sampling of states.
4. Saving in online computation as the states are sampled at different sampling rates compared to the single faster sampling rate.
5. Improvement in the closed-loop performance of the system compared to the single slower sampling rate.
6. Alleviation of system ill-conditioning issues, arising due to multiple time-scales.
7. Improvement in the design accuracy and robustness, as the design is completed in independent design stages.

## 8.2 Future Directions

Finally, a mention is made of the areas that could be pursued for further research but have not been explored in this monograph.

1. For implementing multirate state feedback-based controls, presented in this monograph, continuous-time sequential observers are used. However, the use of multirate full- and reduced-order observers for the two-time-scale and three-time-scale systems could be a promising research area.
2. Development of multi-stage feedback controls for the multi-time-scale systems with time delay is one more research dimension in this area.
3. The formulation of the most convenient models of the three-time-scale linear discrete-time system, i.e., very-fast-time-scale, fast-time-scale, and slow-time-scale formulations, similar to the two-time-scale system and thereafter designing three-stage feedback controllers and multirate feedback controllers.
4. The four-stage controller designs for the four-time-scale systems that have the presence of very slow, slow, fast, and very fast subsystems.

## References

Gajic Z, Lim M-T (2001) Optimal control of singularly perturbed linear systems and applications: high accuracy techniques. Marcel Dekker, New York, USA

Jamshidi M (1983) Large-scale systems: modeling and control. Elsevier, New York

Kokotovic PV, O'Malley RE, Sannuti P (1976) Singular perturbation and order reduction in control theory-an overview. Automatica 12:123–132

Munje RK, Patre BM, Tiwari AP (2018) Investigation of spatial control strategies with application to advanced heavy water reactor. Energy systems in electrical engineering. Springer, Singapore

Naidu DS, Price DB, Hibey JL (1987) Singular perturbations and time-scales (SPaTS) in discrete control systems-an overview. In: Proceedings of the IEEE conference on decision and control 26:2096–2103

Naidu DS (2002) Singular perturbations and time scales in control theory and applications: overview. Dyn Contin, Discret Impuls Syst J 9:233–278

Naidu DS, Rao AK (1985) Application of the singular perturbation method to a steam power system. Electr Power Syst Res 8:219–226

Nguyen HM, Naidu DS (2012) Time scale analysis and control of wind energy conversion systems. In: International symposium on resilient control systems, pp 149–154

Radisavljevic-Gajic V (2015) A simplified two-stage design of linear discrete-time feedback controllers. ASME J Dyn Syst, Meas, Control 137:1–7

Radisavljević-Gajić V, Milanović M, Rose P (2019) Multi-stage and multi-time scale feedback control of linear systems with applications to fuel cells. Mechanical engineering series. Springer, Cham

Saksena VR, O'Reilly J, Kokotovic PV (1984) Singular perturbation and time- scale methods in control theory: survey 1976–1983. Automatica 20:273–293

Shimjith SR, Tiwari AP, Bandyopadhyay B (2013) Modeling and control of a large nuclear reactor a three-time-scale approach. Lecture notes in control and information sciences. Springer Berlin Heidelberg

Syrcos G, Sannuti P (1983) Singular perturbation modeling of continuous and discrete physical systems. Int J Control 37:1007–1022
Zhang Y, Naidu DS, Cai CX, Zou Y (2014) Singular perturbations and time scales in control theories and applications: an overview 2002–2012. Int J Info Syst Sci 9:1–36

The manufacturer's authorised representative in the EU is Springer Nature Customer Service Centre GmbH, Europaplatz 3, 69115 Heidelberg, Germany. If you have any concerns regarding our products, please contact ProductSafety@springernature.com

Printed and bound by CPI Group (UK) Ltd, Croydon, CR0 4YY
30/09/2026
02225309-0009